Hemodynamics in the Echocardiography Laboratory

Gila Perk

Hemodynamics in the Echocardiography Laboratory

Gila Perk
Non Invasive Cardiology
New York Presbyterian-Brooklyn Methodist Hospital
New York, NY
USA

ISBN 978-3-030-79993-9   ISBN 978-3-030-79994-6 (eBook)
https://doi.org/10.1007/978-3-030-79994-6

© The Editor(s) (if applicable) and The Author(s), under exclusive license to Springer Nature Switzerland AG 2021
This work is subject to copyright. All rights are solely and exclusively licensed by the Publisher, whether the whole or part of the material is concerned, specifically the rights of translation, reprinting, reuse of illustrations, recitation, broadcasting, reproduction on microfilms or in any other physical way, and transmission or information storage and retrieval, electronic adaptation, computer software, or by similar or dissimilar methodology now known or hereafter developed.
The use of general descriptive names, registered names, trademarks, service marks, etc. in this publication does not imply, even in the absence of a specific statement, that such names are exempt from the relevant protective laws and regulations and therefore free for general use.
The publisher, the authors, and the editors are safe to assume that the advice and information in this book are believed to be true and accurate at the date of publication. Neither the publisher nor the authors or the editors give a warranty, expressed or implied, with respect to the material contained herein or for any errors or omissions that may have been made. The publisher remains neutral with regard to jurisdictional claims in published maps and institutional affiliations.

This Springer imprint is published by the registered company Springer Nature Switzerland AG
The registered company address is: Gewerbestrasse 11, 6330 Cham, Switzerland

Contents

1 General Principles......1
  1.1 Doppler Principle......2
  1.2 Bernoulli Principle......3
    1.2.1 Bernoulli Equation (Fig. 1.2)......3
  1.3 Wiggers Diagram......5
  1.4 Summary and Final Points......7

2 Math in the Echo Lab......9
  2.1 Volume Calculations......10
    2.1.1 VTI-Based Technique......10
    2.1.2 Border Tracing-Based Technique......12
  2.2 Resistance Calculations......13
  2.3 Continuity Principle......14
  2.4 PISA (Proximal Isovelocity Surface Area) Calculation......15
  2.5 Volumetirc Calculations......18
  2.6 Summary and Final Points......20

3 Pulmonary Pressure: Beginner......21
  3.1 Pulmonary Artery Systolic Pressure......22
  3.2 Analyzing TR Spectral Doppler Envelope......22
  3.3 Measuring Peak TR Velocity......24
  3.4 Assessing Right Atrial Pressure......25
  3.5 Interpreting the Data......25
  3.6 Summary and Final Points......26

4 Pulmonary Pressure: Advanced......27
  4.1 Pulmonary Artery Diastolic Pressure......28
  4.2 Mean Pulmonary Artery Pressure......29
    4.2.1 Calculating Mean PAP Based on Sys/Dias PAP......30
    4.2.2 PI Early Diastolic Velocity......30
    4.2.3 TR VTI......31
    4.2.4 PA Acceleration Time......31

	4.3	Pulmonary Vascular Resistance	32
		4.3.1 Comparing TR ΔP and RVOT Flow	32
		4.3.2 Calculating Pressure Drop and Flow across Pulmonary Circulation	34
	4.4	Summary and Final Points	34
5	**How Severe Is This MR**		37
	5.1	General Principles	38
	5.2	Anatomic/Color Assessment for EROA	40
		5.2.1 3D-Based Direct Measurement of EROA	40
		5.2.2 Vena Contracta Measurement	41
	5.3	PISA Method	42
		5.3.1 Pointers and Potential Pitfalls for the PISA Method	44
	5.4	Color Doppler Assessment	45
	5.5	Volumetric Calculations	46
	5.6	Summary and Final Points	49
6	**What Is Wrong with This MR: Part I**		51
	6.1	Case Presentation	52
	6.2	Approach to an Unusual MR Tracing	53
	6.3	Summary and Final Points	56
7	**What's Wrong with This MR: Part II**		59
	7.1	Case Presentation	60
	7.2	Approach to an Unusual MR Tracing	61
		7.2.1 Low Peak Velocity	61
		7.2.2 Abnormal MR Envelope Shape	61
	7.3	Summary and Final Points	64
8	**Why Is This Happening Now**		65
	8.1	First-Degree Heart Block	66
		8.1.1 Approach to Abnormal MR Timing	66
	8.2	High Degree Heart Block	69
		8.2.1 Approach to Abnormal MR Timing	69
	8.3	The Missing Atrial Contraction	71
		8.3.1 Approach to Abnormal MR Timing	71
	8.4	Summary and Final Points	73
9	**Extreme Pulsus Alternans**		75
	9.1	Case Presentation	76
	9.2	Summary and Final Points	78
10	**Is This AI an Emergency**		79
	10.1	AI Quantification	80
	10.2	AI Hemodynamics	83
	10.3	Summary and Final Points	87

11	**There's a Hole in the Heart: Part I**		89
	11.1	Approach to ASD Evaluation	90
	11.2	Case Presentation	90
		11.2.1 Anatomic Type of ASD	90
		11.2.2 Shunt Direction	91
		11.2.3 Magnitude of the Interatrial Shunt	92
		11.2.4 Effect of the Shunt on Chamber Size and Function	94
		11.2.5 Presence of Pulmonary Hypertension	94
		11.2.6 Estimation of Pulmonary Vascular Resistance	95
	11.3	Case Presentation	96
		11.3.1 Shunt Direction	96
		11.3.2 Magnitude of the Interatrial Shunt	97
		11.3.3 Effect of the Shunt on Chamber Size and Function	97
		11.3.4 Presence of Pulmonary Hypertension	98
		11.3.5 Estimation of Pulmonary Vascular Resistance	98
	11.4	Summary and Final Points	100
12	**There's a Hole in the Heart: Part II**		101
	12.1	Approach to VSD Evaluation	102
	12.2	Case Presentation	102
		12.2.1 Anatomic Type of VSD	103
		12.2.2 Quantifying the Degree of Shunt	103
		12.2.3 VSD Flow and Intracardiac Pressures	104
		12.2.4 Summary of Post MI VSD Case	107
	12.3	Case Presentation	108
		12.3.1 Anatomic Type of VSD	108
		12.3.2 Quantifying the Degree of Shunt	108
		12.3.3 VSD Flow and Intracardiac Pressures	109
		12.3.4 Summary of VSD with Cardiomyopathy Case	110
	12.4	Case Presentation	111
		12.4.1 Anatomic Type of VSD	111
		12.4.2 Quantifying the Degree of Shunt	112
		12.4.3 VSD Flow and Intracardiac Pressures	112
		12.4.4 Summary of VSD with RVOT Obstruction Case	114
	12.5	Summary and Final Points	115
13	**The Machinery Confusion**		117
	13.1	Case Presentation	118
		13.1.1 Approach to PDA Evaluation	118
	13.2	Summary and Final Points	122
14	**All Those AS Gradients**		123
	14.1	Technical Considerations	124
	14.2	Maximal Instantaneous Gradient	125
	14.3	Mean Gradient	126

	14.4	Peak-to-Peak Gradient	127
	14.5	Summary and Final Points	128

15 Dynamic LVOT 129
    15.1 Anatomic Imaging 130
    15.2 Doppler Imaging 132
        15.2.1 Color Doppler Imaging 133
        15.2.2 Spectral Doppler Imaging 133
    15.3 Summary and Final Points 136

16 The Futile Heart 137
    16.1 Case Presentation 138
        16.1.1 Unusual Flow into RA 138
        16.1.2 Quantifying the Shunt 140
    16.2 Summary and Final Points 143

Index 145

General Principles

Abstract

Echocardiography is an important tool in evaluating cardiovascular hemodynamics. There are multiple benefits for utilizing echo for hemodynamics assessment, which include the noninvasive nature of the examination with no known risks, the portable nature of the examination—can be performed at bedside and repeated as needed, and the complementary information that may be obtained and can help in the comprehensive assessment of the hemodynamic status (e.g., chamber size and function, presence of valve disease, and more). Important principles underlying hemodynamic assessment by echocardiography include the Doppler Principle, Bernoulli Principle, and understanding normal intracardiac pressures, flows, and resistance. The Doppler Principle allows the calculation of blood flow velocity by measuring the Doppler shift of a returning signal that encountered a moving target. Blood flow velocity can in turn be used to calculate intracardiac pressure gradients. Understanding relationships between the various intracardiac pressures can help reach conclusions regarding the hemodynamic status.

Keywords

Doppler principle · Doppler equation · Bernoulli principle · Bernoulli equation · Simplified Bernoulli equation · Wigger diagrams · Intracardiac pressures

Introduction

Echocardiography is an important tool in evaluating cardiovascular hemodynamics.

Benefits of utilizing echo for hemodynamics assessment include:
- Noninvasive nature of the examination; no known risks.
- Examination can be done at bedside; no need to transfer unstable patient.

- Ability to repeat the examination if conditions change.
- Available complementary information to help assess hemodynamic status (e.g., chamber size and function, presence of valve disease, etc.).

Important principles underlying hemodynamic assessment by echocardiography include:
- Doppler Principle
- Bernoulli Principle
- Understanding normal intracardiac pressures, flows, and resistance

1.1 Doppler Principle

- Doppler ultrasound (US) is based on scatter interaction between the US wave and red blood cells (Fig. 1.1).
- Since blood cells are a "moving target," the returning US signal has a different frequency than the emitted signal.
- The change in frequency (termed the "Doppler shift"—Δf) is determined by the Doppler equation:
- $\Delta f = \dfrac{2f_0 V}{C} \times \cos\theta$.

f_0—emitted frequency, V—speed of the target, C—propagation velocity of sound in soft tissue, angle θ—Doppler angle of incidence
  - The Doppler angle of incidence is the angle between the US beam and the direction of flow of the "target" (i.e., blood flow).
  - In clinical echocardiography, the US beam is aligned as parallel as possible to the measured flow (angle θ as close to 0^0 as possible → $\cos\theta$ as close to 1 as possible).
  - Attempting to measure and correct for angle of incidence may introduce more error than accuracy.
- The Doppler shift (Δf) is measured by the US machine.

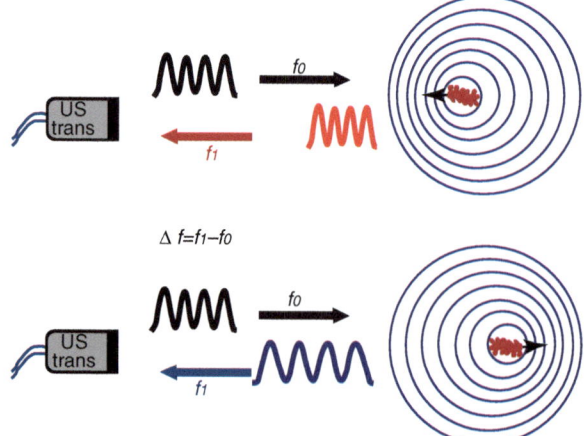

Fig. 1.1 The Doppler effect. When moving blood cells interact with ultrasound, the returning wave has a different frequency (f_1) than the emitted wave (f_0). The change in frequency is the Doppler Shift (Δf), which can be measured by the US machine. The relationship between the Doppler shift and the blood flow velocity is determined by the Doppler equation

1.2 Bernoulli Principle

- Rearranging the Doppler equation to solve for V:

$$V = \frac{\Delta f \times C}{2f_0 (x \cos\theta)}.$$

- Assuming $\cos\theta = 1$.

$$V = \frac{\Delta f \times C}{2f_0}.$$

- By knowing the emitted frequency, propagation velocity of sound in soft tissue, and the measured Doppler shift, blood flow velocity can be calculated.

1.2 Bernoulli Principle

The Bernoulli principle states the following:
Within a horizontal, laminar (streamline) flow, regions with higher fluid speed have lower pressure, and regions with lower fluid speed have higher pressure.
Alternatively, Bernoulli principle can be stated: fluid that flows from high-pressure region to lower pressure region accelerates its flow velocity.

1.2.1 Bernoulli Equation (Fig. 1.2)

- General mathematical way of stating the Bernoulli principle.

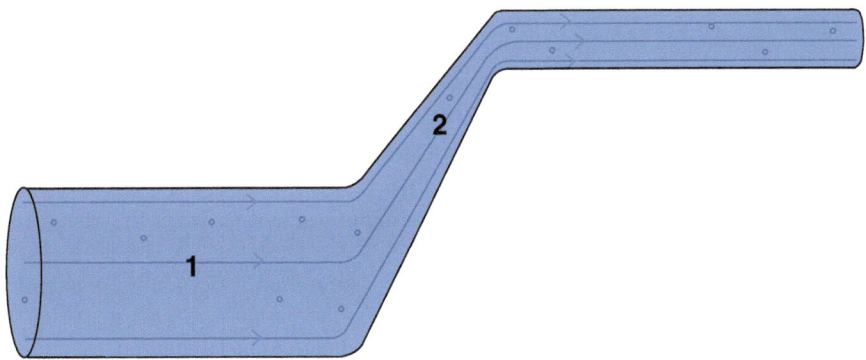

$$P_1 + \frac{1}{2}\rho v_1^2 + \rho g h_1 = P_2 + \frac{1}{2}\rho v_2^2 + \rho g h_2$$

$$P_1 - P_2 = \frac{1}{2}\rho v_2^2 - \frac{1}{2}\rho v_1^2 = \frac{1}{2}\rho(v_2^2 - v_1^2)$$

Blood Density - 1,056 Kg/m³, 1 Pascal = 0.0075 mmHg

$$\Delta P = 4 \times (v_2^2 - v_1^2) = 4v^2$$

Fig. 1.2 Bernoulli principle. Fluid that flows from high-pressure region to lower pressure region accelerates its flow velocity. The Bernoulli equation is the general mathematical way of stating the Bernoulli principle and takes into account changes in kinetic energy, gravitational potential energy, and pressure changes as fluid flows from area 1 to area 2 (P_1/P_2—pressure in Pascal units, v_1/v_2—velocity, h_1/h_2—height at points 1,2 respectively, ρ—fluid density, g—gravity)

- Takes into account changes in kinetic energy and gravitational potential energy as fluid flows from area 1 to area 2:
- $P_1 + \frac{1}{2}\rho v_1^2 + \rho g h_1 = P_2 + \frac{1}{2}\rho v_2^2 + \rho g h_2$.
- $\frac{1}{2}\rho v_1^2$ – Kinetic energy
- $\rho g h_1$ – Potential energy
- P_1-pressure (in Pascal units), v_1-velocity, h_1-height at point 1
- P_2-pressure, v_2-velocity, h_2-height at point 2
- ρ-fluid density, g-gravity
- The above equation can be rearranged as follows:
  P_1-$P_2 = \frac{1}{2}\rho v_2^2 + \rho g h_2 - (\frac{1}{2}\rho v_1^2 + \rho g h_1)$.
- Rearranging the terms on the right:
  P_1-$P_2 = \frac{1}{2}\rho v_2^2 - \frac{1}{2}\rho v_1^2 + \rho g h_2 - \rho g h_1$.
- Assuming no significant change in height between point 1 and point 2: $\rho g h_2 - \rho g h_1 = 0$.
- The equation thus can be simplified to:
  P_1-$P_2 = \frac{1}{2}\rho (v_2^2 - v_1^2)$.
- The density (ρ) of blood is 1056 kg/m³, 1 Pascal = 0.0075 mmHg.
- Rewriting the simplified equation:
  P_1-$P_2 = \frac{1}{2}\rho (v_2^2 - v_1^2) = \frac{1}{2} \times 1056 \times 0.0075 \approx 4 \times (v_2^2 - v_1^2)$ (in mmHg).
- In clinical cardiology, velocity upstream to a narrowing (v_1) is generally low (< 1 m/sec) so the equation can be further simplified:
  $\Delta P = P_1$-$P_2 = 4 \times v_2^2$.

In summary

- The relationship between blood flow velocity and pressure gradient obeys the Bernoulli principle.
- Blood flow velocity is related to the pressure gradient between the chambers where the interrogated flow is happening.
- This relationship is given by the simplified Bernoulli equation: $\Delta P = 4(v_2^2 - v_1^2)$.
- Most commonly the v_1 component (upstream velocity) can be ignored, such that: $\Delta P = 4v^2$.
- Meaning—once blood flow velocity between two chambers is known (calculated by the Doppler equation utilizing scatter interaction between US wave and blood cells) the pressure gradient between the two chambers where this flow occurs can be calculated.

1.3 Wiggers Diagram

- The Wiggers diagram is a standard way to plot intracardiac pressures over time (Fig. 1.3).
  - *X*-axis—Time.
  - *Y*-axis—Ventricular, atrial, and arterial pressures, ECG tracing (±ventricular volume, heart sounds).
- Understanding the relationship between various chambers' pressures and timing of cardiac events (e.g., valve opening / closing), both in normal physiologic conditions and in pathologic states, can help understand normal and abnormal hemodynamics.

Looking at a basic left heart diagram:

Black line—left ventricular (LV) pressure

- Electrical activation sets off ventricular systole.
- Pressure in the LV rises.

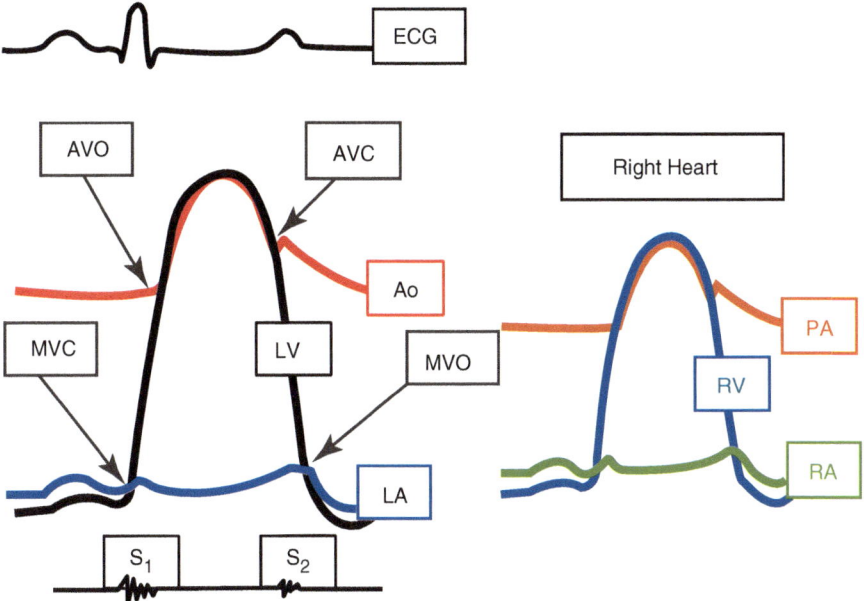

Fig. 1.3 Wiggers diagrams. Standard way to plot intracardiac pressures over time. *X*-axis—time, *Y*-axis—ventricular, atrial and arterial pressures, ECG tracing (Ao—aorta, AVC—aortic valve closure, AVO—aortic valve opening, LA—left atrium, LV—left ventricle, MVC—mitral valve closure, MVO—mitral valve opening, PA—pulmonary artery, RA—right atrium, RV—right ventricle)

- Initially, the pressure rises when both the mitral and the aortic valves are closed such that no change in LV volume occurs; this period is the isovolumetric contraction time—IVCT.
- Once LV pressure rises above aortic pressure, the aortic valve opens; ejection period starts.
- LV pressure starts to drop at end systole; once LV pressure drops below aortic pressure, the aortic valve closes.
- LV relaxation continues with both mitral and aortic valves closed; this period is the isovolumetric relaxation time—IVRT.
- Once LV pressure drops below LA pressure, the mitral valve opens and rapid filling of the LV begins.

Blue line—left atrial (LA) pressure

- Late diastolic/pre-systolic LA pressure increase is associated with the atrial contraction.
- Following atrial contraction, atrial pressure drops due to atrial relaxation.
- When LA pressure drops below LV pressure (due both to atrial relaxation and to start of ventricular systole), the mitral valve closes.
- Slight increase in atrial pressure is noted as the mitral valve closes.
- Left atrial pressure gradually increases during ventricular systole due to filling from the pulmonary veins.
- Ventricular pressure drops during LV relaxation; when LV pressure drops below LA pressure, the mitral valve opens.
- Following mitral valve opening, rapid filling of the LV from the LA starts.

Red line—aortic (Ao) pressure

- Diastolic aortic pressure is higher than LV pressure.
- As ventricular systole starts, the LV pressure increases until it exceeds the Ao pressure, at which point the aortic valve opens.
- Under normal conditions, the pressure in the aorta is nearly equal to the pressure in the LV during systole.
- Upon end of LV contraction, pressure in the LV drops; when the pressure falls below aortic pressure, the aortic valve closes.
- A dicrotic notch can be seen on Ao pressure tracing at the time of aortic valve closure.

Similar diagrams depict intracardiac right heart pressures; absolute values of peak and trough pressures differ from the left heart and duration of IVCT/IVRT can differ, however the general principles are essentially similar.

1.4 Summary and Final Points

- The Doppler principle allows calculation of blood flow velocity by measuring the Doppler shift of a returning signal that encountered a moving target.
- Note:
  - When measuring blood flow velocity, the US beam is aligned as parallel as possible to the interrogated flow. No angle correction is used in cardiac ultrasound.
  - If the US beam is not within 20–30° of the interrogated flow, the calculated flow velocity will be underestimated.
  - When used correctly, Doppler-based velocity **cannot** overestimate true velocity.
  - When a specific blood flow is interrogated from multiple views/angles, the highest velocity is the most accurate one, as it reflects the most parallel acquisition.
  - However, if there are **physiologic** reasons for variability in velocity (e.g., irregular heart rate with variable R-R intervals, respiratory variations), averaging velocities from several beats is required. In these cases, variability will appear in tracings obtained from the same view (rather than variability that is obtained by recording velocities from different windows/angles).
- In the vast majority of circumstances, blood flow velocity is determined by the pressure gradient driving the measured flow.
- Thus, blood flow velocity can be used to calculate intracardiac pressures.
- The pressure gradient driving the measured flow is calculated using the simplified Bernoulli formula ($\Delta P = 4v^2$).
- Whenever blood velocity is measured, the following questions need to be asked:
  - Between what two chambers does the flow occur?
  - At what part of the cardiac cycle?
- In most circumstances, blood flow velocity is **not** related to quantity; for instance, mitral regurgitation peak velocity is **not** a manifestation of MR severity, it **is** a manifestation of the pressure gradient between the left ventricle and the left atrium during systole.
- In most cases, it is helpful to conceptually separate assessment of velocity and quantity; velocity measurements provide information about pressures, volumetric calculations (see Chap. 2) provide information about quantity.
- In rare circumstance, the Bernoulli principle cannot be applied to the interrogated flow; examples include nonrestrictive flow, serial obstructions, and more.
- Once the pressure gradient is calculated, utilizing the Wiggers diagrams and the known relationships between the various intracardiac pressure can help reach conclusions regarding the hemodynamic status.

Math in the Echo Lab

Abstract

Several calculations are utilized in echocardiography to help assess hemodynamic status. Many of these calculations can be carried out at various locations in the heart such that different volumes and pressures can be estimated. In order to understand the clinical significance of the obtained parameters, all available information needs to be taken into consideration. Understanding possible shortcomings of the techniques, impact of technical limitations, and the effects of various cardiovascular pathologies on these parameters is paramount for proper interpretation of the obtained data. The basic parameters and principles that are utilized when assessing hemodynamics by echocardiography include: volume calculations, resistance calculation, Continuity principle, PISA method, and volumetric calculations.

Keywords

VTI · Volume calculations · Resistance calculation · Continuity principle · PISA method · Volumetric calculations · Simpson's rule · Method of discs · Stroke volume · Cardiac output · Systemic vascular resistance · Pulmonary vascular resistance

Introduction

- Several calculations are utilized in echocardiography to help assess hemodynamic status.
- Many of these calculations can be carried out at various locations in the heart.
- In order to understand the clinical significance of the obtained parameters, all available information needs to be taken into consideration.

- Understanding possible shortcomings of the techniques, impact of technical limitations, and the effects of various cardiovascular pathologies on these parameters is paramount for proper interpretation of the obtained data.

In this chapter, we will review the basic parameters and principles that are utilized when assessing hemodynamics by echocardiography
- Volume calculations
- Resistance calculation
- Continuity principle
- PISA method
- Volumetric calculations:

2.1 Volume Calculations

2.1.1 VTI-Based Technique

- Doppler echocardiography can be utilized to measure flow volumes.
- Spectral Doppler tracings plot flow velocity against time (Fig. 2.1)
    - X-axis—time
    - Y-axis—flow velocity
    - By convention, velocity toward the transducer is depicted above the baseline, flow away from the transducer is shown below the baseline.
- Tracing the spectral Doppler signal over time (one beat) allows measurement of the velocity time integral —VTI.

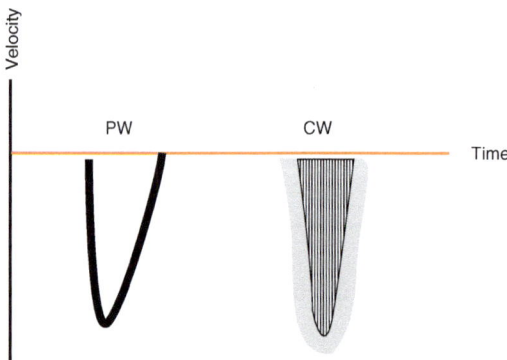

Fig. 2.1 Spectral Doppler. Spectral Doppler tracings plot flow velocity against time; X-axis—time, Y-axis—flow velocity. By convention, velocity toward the transducer is depicted above the baseline, flow away from the transducer is shown below the baseline. Pulse-wave (PW) Doppler tracings record velocity from one location creating a spectral envelope with outline only. Continuous-wave (CW) Doppler tracings record velocities from anywhere along the path of the US beam creating a full envelope

Fig. 2.2 VTI. Tracing a spectral Doppler envelope yields the VTI. The VTI can be thought of as representing the "height" (*h*) of the column of blood that passed through the site where it was measured, during one beat. (Ao—aorta, CSA—cross-sectional area, Desc Ao—descending aorta, LA—left atrium, LV—left ventricle)

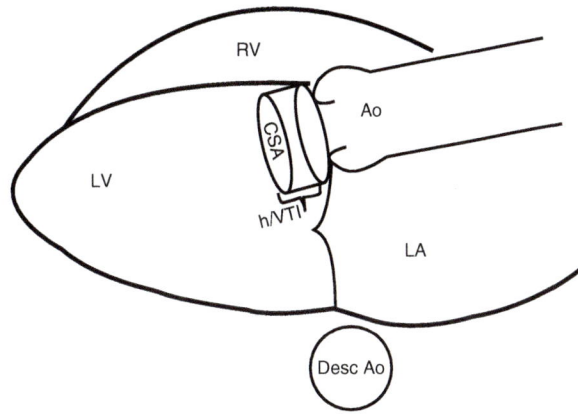

- The VTI can be thought of as representing the "height"' of the column of blood that passed through the site where it was measured, during one beat (Fig. 2.2).
  - Note: When integrating velocity ($\frac{distance}{time}$) over time, the obtained number has distance units; in echocardiography, VTI is measured in centimeters.
- If the cross-sectional area at the site of the measured flow is known, the volume of blood that passes through this site can be calculated (utilizing cylinder volume calculation formula):
  $V = CSA \times VTI$
  V—Volume
  CSA—Cross-sectional area at the site of the measured flow
  VTI—velocity time integral at the site

Since the VTI is obtained by tracing one cycle, the volume obtained by this method is the stroke volume (SV).

Important points to remember:
- This technique can be utilized at any cardiac location where CSA and spectral Doppler can both be measured (examples include left ventricular outflow tract (LVOT), mitral annulus, right ventricular outflow tract).
- CSA is often obtained by measuring diameter and calculating the area by circle-area formula, or occasionally ellipse-area formula (two orthogonal diameters are required).
- When feasible, direct measurement of CSA can also be used (e.g., utilizing 3-dimensional acquisition with post-processing).
- CSA/diameter and VTI should be obtained at the same phase of the cardiac cycle; meaning if LVOT is utilized to measure stroke volume, VTI is obtained during systole and hence diameter measurement should be performed on a systolic frame.

- In normal cardiac physiology, stroke volume is equal across all chambers and valves; however, when regurgitant lesions or intracardiac shunts are present, volumes calculated across different valves can differ. It should be clearly understood what the calculated volumes represent (e.g., including regurgitant volume or true forward stroke volume).
- Once stroke volume is known, cardiac output (CO) can be calculated by:
  CO = SV × HR
  HR—heart rate

2.1.2 Border Tracing-Based Technique

- 2D images can be utilized to measure volumes.
- Adequate image quality is essential for the accuracy of this technique. Attention should be paid to:
  - Adequate endocardial definition with clear delineation of endocardial–blood border; if needed, Ultrasound Enhancing Agents (UEA), "echo contrast," may be used to improve endocardial resolution.
  - Avoidance of foreshortening of the chamber.
  - Proper timing of the frames utilized for tracings.
- The most commonly used method for volume calculations based on endocardial border tracing is the Simpson's Rule/Method of discs (Fig. 2.3).
- For LV volume assessment, endocardial border is traced at end diastole and at end systole from two orthogonal views (apical 4-chamber view and apical 2-chamber view).
  - End diastole is defined as the frame immediately following mitral valve closure; if this cannot be accurately discerned, the frame that occurs with the peak of the QRS complex is chosen.
  - End systole is defined as the frame with the smallest LV volume visualized.
- For LA volume assessment, endocardial border is traced at maximal LA dimension (before mitral valve opening) from two orthogonal views (apical 4-chamber view and apical 2-chamber view).
- The traced area is divided into a series of "stacked discs" (usually 20).
- The volume of each disc is calculated as a cylinder volume.
  - When two orthogonal views are used, the radius (r) of each cylinder is measured from each view; volume of the disc is then calculated by: $V = \pi r_1 r_2 \times h$.
- When LV is traced, end diastolic and end systolic volumes (EDV, ESV) are obtained.
- Stroke volume (SV) is calculated as: SV = EDV − ESV.
- Ejection fraction is calculated as: $\mathrm{EF} = \dfrac{\mathrm{EDV} - \mathrm{ESV}}{\mathrm{EDV}} \times 100 = \dfrac{\mathrm{SV}}{\mathrm{EDV}} \times 100$.
- Important to note that this technique measures "total" stroke volume; if a regurgitant lesion is present (i.e., mitral or aortic regurgitation), the calculated stroke volume **includes** the regurgitant volume and does **not** represent the forward (or "effective") stroke volume.

2.2 Resistance Calculations

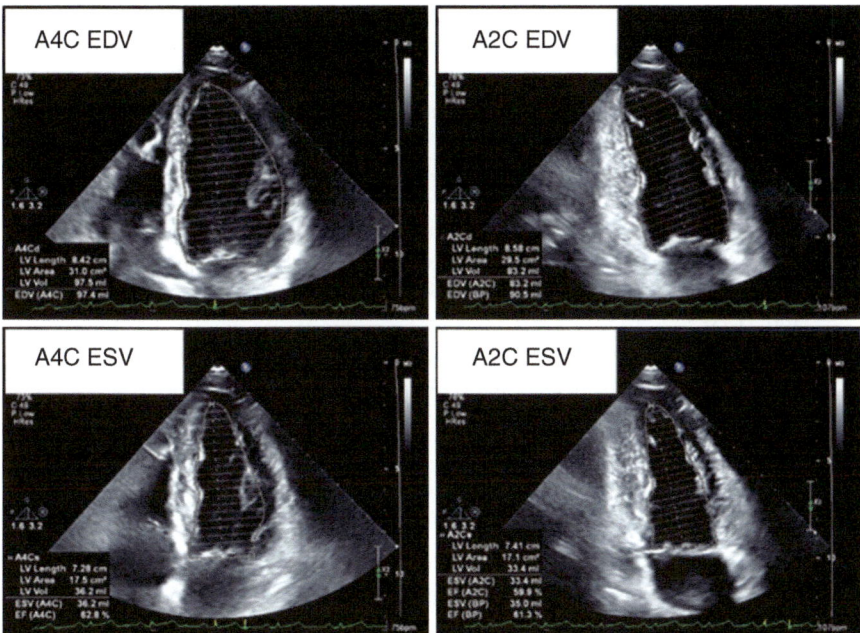

Fig. 2.3 Method of discs. Endocardial border is traced at end diastole and at end systole from two orthogonal views (apical 4—chamber view and apical 2—chamber view). The traced area is divided into a series of "stacked discs" and the volume of each disc is calculated as a cylinder volume. When LV is traced, end diastolic and end systolic volume (EDV, ESV) are obtained

2.2 Resistance Calculations

- Systemic and pulmonary vascular resistances can often be calculated by echocardiography.
- The calculation is based on the relationship between flow (F), pressure difference (ΔP), and resistance (R); the flow is directly related to the pressure difference across the system and inversely related to the resistance within the system: $F = \dfrac{\Delta P}{R}$.
- The above relationship is important to remember when evaluating any increased pressure gradient; since $\Delta P = F \times R$, high-pressure gradient (ΔP) can be the result of increased flow (F) or increased resistance (R).
- Differentiating between these causes is extremely important, prognostically and therapeutically.
- Rearranging the above equation: $R = \dfrac{\Delta P}{F}$.
- This formula can be used to calculate systemic vascular resistance (SVR) and pulmonary vascular resistance (PVR).

- For calculating SVR, the pressure drop (ΔP) across the systemic circulation needs to be calculated.
  - Systemic ΔP = Mean arterial pressure (MAP) – Right atrial pressure (RAP).
  - The flow across the systemic circulation equals the systemic cardiac output.
  - $SVR = \dfrac{MAP - RAP}{CO}$ (Woods units).
- For calculating PVR, the pressure drop across the pulmonary circulation needs to be calculated:
  - Pulmonary ΔP = Mean pulmonary artery pressure (MPAP) – Left atrial pressure (LAP).
  - Under normal conditions (no shunt), the flow across the pulmonary circulation equals the flow across the systemic circulation (which equals the cardiac output).
  - $PVR = \dfrac{MPAP - LAP}{CO}$.

Important points to remember:
- Mean arterial pressure (and not *systolic* pressure) should be used to calculate ΔP across the circulatory system (mean arterial pressure or mean pulmonary artery pressure).
- When pulmonary artery pressure is assessed by echo, typically the pulmonary artery systolic pressure is estimated (see Chap. 3 for discussion regarding pulmonary artery pressure assessment); in order to calculate PVR, additional information should be obtained such that mean PAP can be estimated.
- When significant intracardiac shunts are present, systemic cardiac output does not equal pulmonary cardiac output. Right heart and left heart cardiac outputs should be measured separately and used accordingly in the resistance formula.

2.3 Continuity Principle

- Looking at a flow through a tube that narrows (Fig. 2.4), the continuity principle dictates that flow velocity (v) must increase as the fluid passes through a narrow part of the tube.
- The continuity principle is based on conservation of mass; the amount of fluid that enters the tube at the wide end (A_1) during a period of time (T), equals the amount of fluid that leaves the tube at the narrow end (A_2) during that same period of time (T).
- The flow rate (Q) is constant; $Q = \dfrac{\text{Volume}}{\text{Time}}$.
- The volume of fluid the enters (or leaves) the tube can be calculated as the volume of a cylinder: *Volume = Cross Sectional Area (CSA) × Distance (d)*.
- From above equation

2.4 PISA (Proximal Isovelocity Surface Area) Calculation

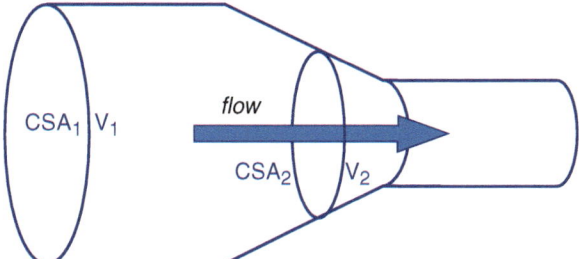

Fig. 2.4 Continuity principle. The continuity principle, which is based on conservation of mass, dictates that flow velocity (v) must increase as the fluid passes through a narrow part of the tube. Since the amount of fluid that enters the tube at a wide end (CSA_1) during a period of time (T), equals the amount of fluid that leaves the tube at a narrow end (CSA_2) during that same period of time (T), flow velocity (v) must accelerate

- $Q = \dfrac{\text{Volume}}{\text{Time}} = \dfrac{CSA \times d}{\text{Time}} = CSA \times v$ (v = flow velocity of the fluid).
- The continuity principle states that the volume that enters the tube at A_1 must equal the volume of fluid that leaves at A_2
- $CSA_1 \times v_1 = CSA_2 \times v_2$.
- Since CSA_2 is smaller, v_2 must be larger; meaning flow velocity accelerates as the tube narrows.
- In echocardiography, the continuity principle is used to calculate the unknown area.
- Rearranging the above equation:
- $CSA_2 = \dfrac{CSA_1 \times v_1}{v_2}$
- If one area is known (e.g., CSA_1-LVOT cross-sectional area) and flow velocity can be measured at the LVOT (v_1) as well as flow velocity across the aortic valve (v_2), the equation can be used to calculate aortic valve area (CSA_2).

2.4 PISA (Proximal Isovelocity Surface Area) Calculation

PISA method can be used to calculate:

- Effective regurgitant orifice area (EROA).
- Mitral valve area (MVA); although significant limitations exist for using PISA method for this purpose, the principles still apply.

PISA method is based on the following principles:

- Continuity principle: The volume of blood flowing immediately proximal to the valve (or proximal to a regurgitant orifice) is equal to the flow through the valve (or through the EROA).
- As blood flows toward a smaller orifice (EROA or tips of the mitral valve), the flow converges to form concentric hemispheres, until ultimately it passes through the EROA/valve tips.
- Color Doppler imaging does not allow direct measurement of flow velocity (only average velocities are depicted by the color); however, aliasing of the color jet (the point where color changes from laminar blue/red flow to mosaic/turbulent yellowish flow) occurs at a known velocity, which is shown on the color scale of the image (the Nyquist limit).
- Maximum velocity of the flow occurs at the smallest orifice along the way, which is the EROA for regurgitant lesions, or the tips of the mitral valve (for forward flow through the valve).

Applying the above principles (Fig. 2.5):
- As the regurgitant flow accelerates toward the regurgitant orifice, concentric hemispheres are formed, where flow velocities at the surface of each hemisphere are equal.
- The hemisphere that can be easily identified is the one where blood flow velocity reaches precisely the Nyquist limit; this hemisphere is identified by the change of the color signal from laminar (red or blue flow) to the mosaic/yellow flow.

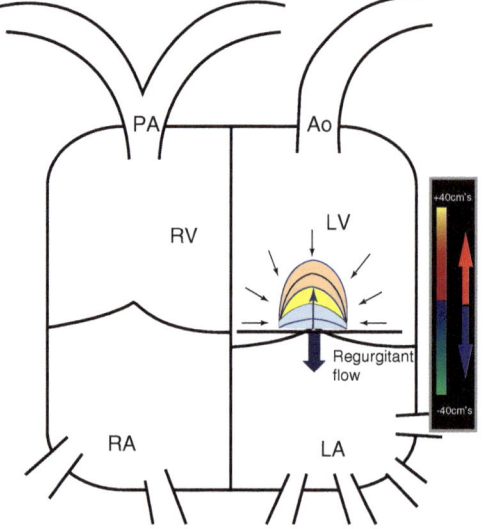

Fig. 2.5 PISA method. As the regurgitant flow accelerates toward the regurgitant orifice, concentric hemispheres are formed, where flow velocities at the surface of each hemisphere are equal. The hemisphere where blood flow velocity reaches the Nyquist limit can be easily identified and its radius (r) measured (Ao—aorta, LA—left atrium, LV—left ventricle, PA—pulmonary artery, RA—right atrium, RV—right ventricle)

2.4 PISA (Proximal Isovelocity Surface Area) Calculation

- Once the hemisphere is identified, its radius (r) can be measured as the distance from the surface (where the color changes) to the point of valve leaflets coaptation.
- The surface area (SA) of the hemisphere is calculated by SA = $2\pi r^2$.
- The flow velocity at the surface of this hemisphere is known—it is the Nyquist limit/aliasing velocity (V_{alias}) which can be seen on the scale at the corner of the image.
- Flow at the surface of the hemisphere can thus be calculated Q = SA × V_{alias} = $2\pi r^2$ × V_{alias}.
- Since the regurgitant orifice is the smallest area along the path, the jet reaches maximum velocity precisely at the regurgitant orifice.
- The flow through the regurgitant orifice can be calculated as Q = EROA × V_{max}.
- By the continuity principle, the flow at the surface of the hemisphere equals the flow through the regurgitant orifice: $2\pi r^2$ × V_{alias} = EROA × V_{max}.
- Rearranging the equation: EROA = $\dfrac{2\pi r^2 \times V_{alias}}{V_{max}}$.

Important points to remember:
- The EROA is directly related to the measured hemisphere radius (r); the larger the radius, the larger the EROA (the worse the regurgitant lesion).
- If using the PISA method to calculate mitral valve area, the measured r is still directly related to the calculated area (MVA).
- Thus, larger measured r → larger MVA.
- When assessing mitral stenosis (MS) severity—larger MVA implies **less** severe MS.
- Hence, when using PISA method for MVA assessment: larger PISA r → less severe MS.
- Limitations on using PISA for MVA calculation:
  - When the mitral valve opens, it forms a "funnel shape"; the accelerated flow does not form a full hemisphere but rather a section of a hemisphere (a "three-dimensional pizza slice").
  - In order to calculate the surface area of the hemisphere section that is formed, a correction for the angle between the mitral valve leaflets (angle θ) should be used.
  - The section surface area (SSA) is calculated as SSA = $2\pi r^2 \times \dfrac{\theta}{180}$.
  - The flow velocity through the mitral valve changes throughout diastole; it depends on mitral valve area as well as other factors (e.g., LV compliance and atrial contraction).
  - Choosing the velocity to use in the MVA formula may be more challenging and less obvious than when using PISA method for regurgitant orifice calculation.

2.5 Volumetirc Calculations

- As mentioned above, the continuity principle is based on conservation of mass; the volume of fluid that enters a system at one end (CSA_1) during a period of time (T), equals the volume of fluid that leaves the system at the other end (CSA_2) during that same period of time (T).
- Under normal circumstances (e.g., no regurgitant lesions or shunts), the volume that flows across any cardiac valve during one cycle, should equal the flow volume across any other cardiac valve (or area, i.e., LVOT/RVOT).
- Any discrepancy between calculated flow volumes should be accounted for and explained.
- There can be technical reasons for calculated volume differences, as well as pathophysiologic reasons; it is of utmost importance to distinguish between these possible causes.
    - Technical issues produce mistakes in the calculations (preventable or non-preventable).
        Examples include:
        Inaccuracy in diameter measurement due to limited spatial resolution of the echocardiographic image.
        Erroneous cross-area calculation due to assumption of a round area configuration when a true area may have a different shape (e.g., mitral annulus is saddle shaped—formally a "Hyperbolic Paraboloid" or more easily—a Pringles Chip shape; assuming a round configuration results in an error in the calculated mitral annular area).
        Improper positioning of the pulse-wave Doppler sample volume to measure the VTI across the desired area.
    - Pathophysiologic issues are situations in which the flow volumes across the various valves/chambers do in fact differ. In these cases, assuming no measurement errors are made, volumetric calculations can provide quantitative information regarding the lesion being investigated:
        Example (Fig. 2.6):
        When mitral regurgitation (MR) is present, comparing the inflow across the mitral annulus and the outflow across the LVOT can provide a quantitative assessment of the MR severity.
        The inflow across the mitral annulus is calculated by measuring the annular diameter (d) and the mitral annulus VTI (measured by pulse-wave Doppler with sample volume placed at the mitral annulus); for accurate results, mitral annulus diameter should be measured in two planes (from parasternal long-axis view and from apical four-chamber view).
        Annular area is calculated using an ellipse area formula:
        Annular CSA = $\pi MVr_1 MVr_2$ (note: $MVr = \dfrac{d}{2}$).
        Inflow volume across the mitral annulus is calculated as:
        V_{inflow} = Annular CSA × Annular VTI.

2.5 Volumetric Calculations

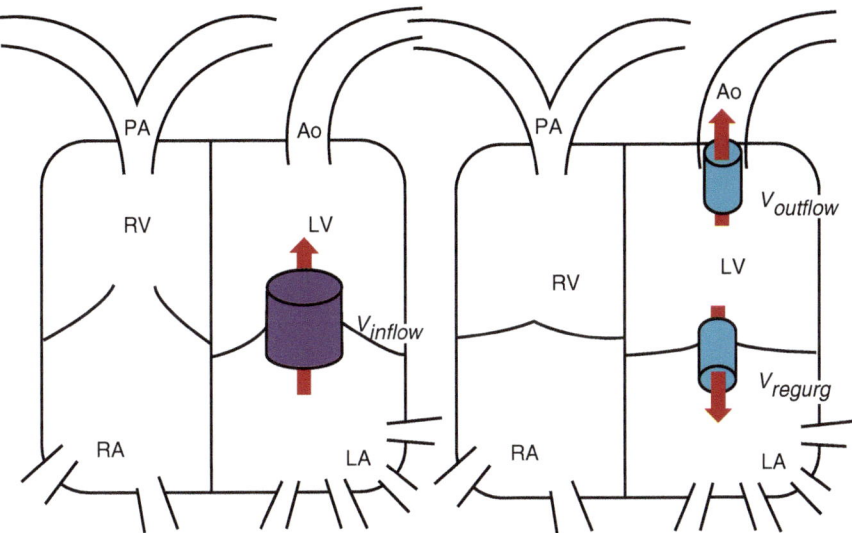

Fig. 2.6 Volumetric calculation. When mitral regurgitation (MR) is present, comparing the inflow across the mitral annulus (V_{inflow}) and the outflow across the LVOT ($V_{outflow}$) can provide a quantitative assessment of the MR regurgitant volume (V_{regurg}). (Ao—aorta, LA—left atrium, LV—left ventricle, PA—pulmonary artery, RA—right atrium, RV—right ventricle)

The LV outflow is calculated by measuring the LVOT diameter and the LVOT VTI (measured by pulse-wave Doppler with sample volume placed at the LVOT); careful attention should be paid to accurate LVOT diameter measurement.

LVOT area is calculated as: LVOT CSA $=\pi LVOT r^2$.

Outflow across the LVOT is calculated as: $V_{outflow}$=LVOT CSA × LVOT VTI. In the presence of mitral regurgitation, the inflow across the mitral annulus **does not** equal the outflow across the LVOT; the mitral inflow **includes** the regurgitant volume, whereas the LVOT outflow **does not**.

Comparing these two volumes allows calculation of the mitral regurgitant volume (RV) as the difference between the mitral annular inflow and the LVOT outflow

$RV = V_{inflow} - V_{outflow}$
   = Annular CSA × Annular VTI − LVOT CSA × LVOT VTI.

Important points to remember when using volumetric techniques for quantification:

- Since volumes are compared across two different valves/cardiac areas, volumetric method can only be used in the presence of a single lesion (e.g., MR only); if more than an isolated pathology is present (e.g., mitral regurgitation and aortic regurgitation), the difference in calculated volumes may be multifactorial.
- Errors in diameter/radius measurements can produce significant errors in volume calculations since these measurements are squared when calculating CSA.

- Heart rate variability may result in beat-to-beat variability in stroke volume; a possible solution is to average several beats' VTIs for flow-volume calculation.

2.6 Summary and Final Points

- Blood flow velocity is calculated by utilizing the Doppler principle.
- Most commonly, the blood flow velocity is determined by the pressure gradient driving the flow; thus, measuring blood flow velocity allows calculation of intracardiac pressure gradients.
- Integrating velocity over time gives the VTI—the velocity time integral—which is a measure of the height of the column of blood that passes through a certain area during the measured time.
- Measuring VTI, at a site that can be measured for its cross-sectional area (CSA), provides a means to calculate the volume of blood that passes through that area at a given time.
- Under normal circumstances, the volume that flows across any cardiac valve or any cardiac area (e.g., RVOT/LVOT) during one beat is equal between all sites—it is the stroke volume.
- When regurgitant lesions or shunts are present, comparing the calculated volumes, obtained from measurements at two different sites, can help assess the severity of the interrogated lesion.
- Care must be taken to avoid technical errors (e.g., inaccurate diameter measurements and wrong sample volume placement) as well as conceptual errors (e.g., multivalve disease with more than a single regurgitant volume).
- All available data should be analyzed and interpreted in conjunction such that reasonable conclusions can be drawn.

Pulmonary Pressure: Beginner

Abstract

Echocardiography can be utilized to assess pulmonary pressures. Pulmonary pressures are an important part of the hemodynamic assessment and carry significant prognostic and therapeutic implications. The benefits of using echocardiography for assessing pulmonary pressures include the noninvasive nature of the assessment with no known risks, the ability to perform the study at bedside and repeat as needed, and the availability of complementary information (e.g., right ventricular size and function) that can help understand that pathologic process and its impact on the heart. It is essential to understand which pulmonary pressure (e.g., systolic, diastolic, or mean) is measured by each technique in order to have a comprehensive assessment, as well as compare data across various techniques. The most commonly used measurement to assess pulmonary pressure is the interrogation of tricuspid regurgitation jet velocity. This allows for calculation of the pulmonary artery systolic pressure, which provides a good estimate for the presence and severity of pulmonary hypertension.

Keywords

Tricuspid regurgitation · Pulmonary artery systolic pressure · Right atrial pressure

Introduction

Echocardiography can be frequently utilized to assess pulmonary pressures. Pulmonary pressures are an important part of the hemodynamic assessment and carry significant prognostic and therapeutic implications.

The benefits of using echocardiography for assessing pulmonary pressures:
- Noninvasive nature of the assessment; no known risks.
- Measurements can be done at bedside; no need to transfer an unstable patient.

- Assessment can be repeated if conditions change.
- Availability of complementary information (e.g., right ventricular size and function).

It is essential to understand which pulmonary pressure (e.g., systolic, diastolic, or mean) is measured by each technique in order to have a comprehensive assessment, as well as compare data across various techniques.

In this chapter, we will review the most commonly used measurement to assess pulmonary artery systolic pressure. Chapter 4 will review more advanced techniques to assess pulmonary artery diastolic pressure, mean pulmonary artery pressure, and pulmonary vascular resistance.

3.1 Pulmonary Artery Systolic Pressure

- Trace-to-mild tricuspid regurgitation (TR) is seen in the majority of healthy individuals.
- It is considered a normal variant; in and of itself does not require any follow-up or treatment.
- Measuring the TR peak velocity can provide information regarding pulmonary artery systolic pressure.
- Since trace-mild TR is so common, this measurement is available in nearly all echocardiograms.

3.2 Analyzing TR Spectral Doppler Envelope

- TR velocity should be assessed from multiple windows and views.
- Continuous-wave (CW) Doppler is used to record the TR tracing.
- Quality of the spectral tracing should be evaluated; full envelope should be seen, peak velocity clearly demarcated.
- The density of the TR spectral tracing (e.g., how white/bright the TR tracing is), is **not** determined by the pressure gradient, but rather by the quantity of the TR.
- If different velocities are recorded from different views, the velocity from the window that provides the highest measurement should be used, as this represents the most parallel acquisition of the TR spectral Doppler.
- If beat to beat variability exists (e.g., irregular heart rate) or respiratory variations are present, averaging several beats (from the highest velocity window) should be done.
- Typically TR envelope is parabolic shaped, mid-peaking (Fig. 3.1).
- When severe/torrential TR is present with poor coaptation of the valve leaflets, the TR envelope appears triangular and early peaking; there is early equilibration of pressures between the right ventricle (RV) and the right atrium (RA) (Fig. 3.2).

3.2 Analyzing TR Spectral Doppler Envelope 23

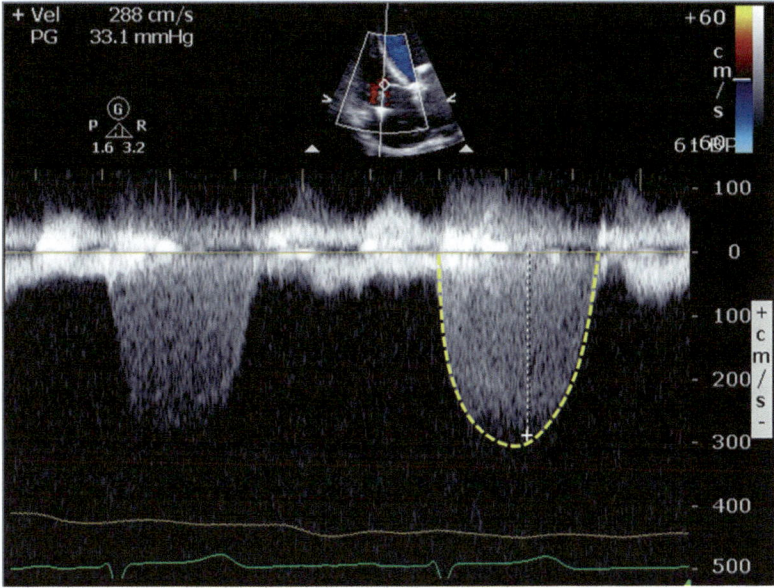

Fig. 3.1 Tricuspid regurgitation. Typical TR spectral Doppler has a parabolic, mid-peaking shape

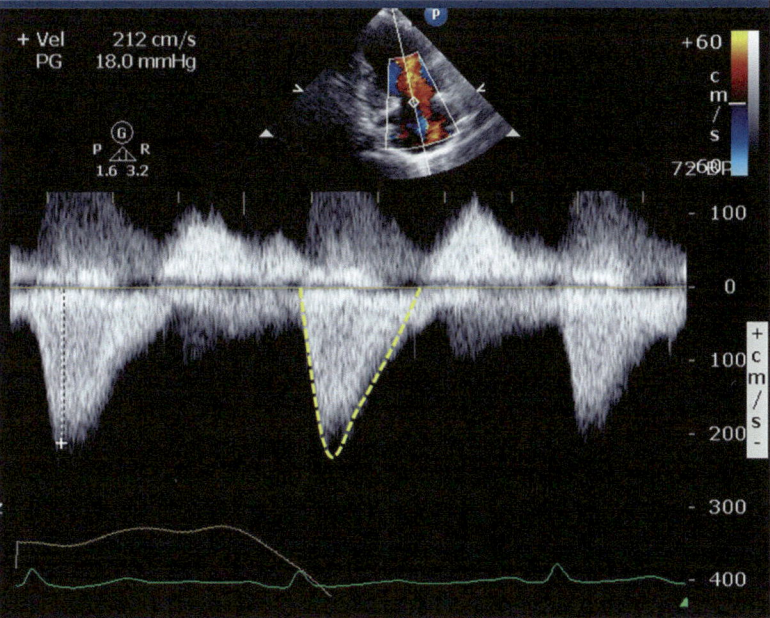

Fig. 3.2 Torrential tricuspid regurgitation. When there is severe TR due to non-coapting tricuspid leaflets, the TR envelopes appear triangular and early peaking. There is rapid equilibration of the right ventricular and right atrial pressure in early systole

- In cases with torrential TR and triangular-shaped TR envelope, the assessment of PA systolic pressure, utilizing peak TR velocity becomes inaccurate (see below) and should not be reported.

3.3 Measuring Peak TR Velocity

- When TR envelope is suitable for analysis, peak TR velocity is measured (Fig. 3.3).
- Using the simplified Bernoulli equation, the pressure gradient driving the TR is calculated as $\Delta P = 4v^2$.
- The calculated pressure gradient is the difference between the systolic RV pressure and the systolic RA pressure (RAP).
- Thus, the right ventricular systolic pressure can be calculated as: RV SysP = RAP + ΔP.
- Looking at the Wiggers diagrams of the right heart—during systole, when the pulmonic valve is open (and no pulmonic stenosis), the RV systolic pressure equals the pulmonary artery systolic pressure.
- Thus, the PA systolic pressure is: PA SysP = RAP + ΔP.

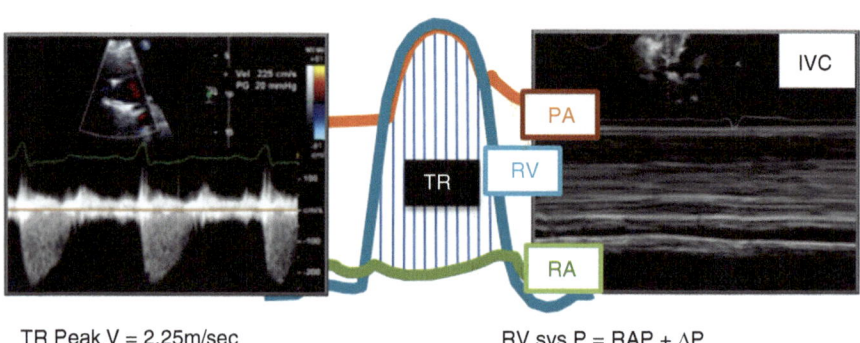

TR Peak V = 2.25m/sec

RV sys P = RAP + ΔP
PA sys P = RV sys P

Fig. 3.3 Systolic pulmonary artery pressure. Peak tricuspid regurgitation (TR) velocity is measured on an adequate TR spectral envelope. The pressure gradient driving the TR is calculated as $\Delta P = 4v^2$ and represents the difference between the systolic right ventricular (RV) pressure and the systolic right atrial (RA) pressure. In the absence of pulmonic stenosis (rare) the RV systolic pressure equals the pulmonary artery systolic pressure; thus, the PA systolic pressure can be estimated as: PA SysP = RAP+ ΔP

3.4 Assessing Right Atrial Pressure

- The right atrial pressure (RAP) is estimated by looking at the inferior vena cava (IVC) from the subcostal view; IVC size and inspiratory collapse are assessed.
    - IVC size is measured slightly proximal to the IVC–RA junction (approximately 1–2 cm proximal to the junction).
    - Inspiratory change can be measured during quiet respiration; if needed, the patient can be asked to perform a "sniff." It is important to make sure that the image does not "disappear" during the sniff test and actual change in IVC diameter is observed.
    - RA pressure is estimated as normal (~3 mmHg) if IVC diameter is less than 2.1 cm with greater than 50% inspiratory collapse.
    - RA pressure is estimated as elevated (` ~ 15 mmHg or higher) if IVC diameter is greater than 2.1 cm with less than 50% inspiratory collapse.
    - RA pressure is estimated as mildly elevated (~7–8 mmHg) with in-between parameters (IVC diameter less than 2.1 cm but lower than 50% inspiratory collapse or IVC diameter greater than 2.1 cm with normal inspiratory collapse).
    - Notably, young healthy individuals may have IVC diameter greater than 2.1 cm; this should not be interpreted as elevated RA pressure.
    - RA pressure estimate by IVC evaluation should always "make sense"—it should be interpreted in light of other parameters like RA size and TR severity.

3.5 Interpreting the Data

- Utilizing peak TR velocity to assess pulmonary artery systolic pressure is the most commonly used technique.
- Important to remember that the TR technique for PA pressure provides an estimation of the PA **systolic** pressure.
- The WHO definitions of presence and severity of pulmonary hypertension relay on **mean** pulmonary artery pressure.
- Systole accounts for approximately a third of the cardiac cycle; severity of systolic pulmonary hypertension may not necessarily reflect the true severity of the pulmonary hypertension.
- Given that, echocardiographic cutoffs for presence and severity of pulmonary hypertension differ from those provided by the WHO.
- Exact echocardiographic cutoffs are of some debate and differ across various laboratories.
- In addition, pulmonary pressure generally increases with age, such that cutoffs might be different for different age groups.

- Example of a commonly used gradation system:
  - Pulmonary hypertension present: PA systolic pressure estimated to be >40 mmHg.
  - Mild pulmonary hypertension: PA systolic pressure between 40 and 50/55 mmHg.
  - Moderate pulmonary hypertension: PA systolic pressure between 50–55 and 60/65 mmHg.
  - Severe pulmonary hypertension: PA systolic pressure > 65 mmHg.

3.6 Summary and Final Points

- Trace-to-mild TR is extremely common and can be used to estimate pulmonary artery systolic pressure.
- Signal quality (on the CW spectral tracing) must be adequate for proper estimation.
- If the signal is inadequate, reimaging with agitated saline injection or contrast injection may enhance the signal and allow for better delineation of the spectral envelope.
- If peak TR velocity cannot be accurately discerned, PA pressure should not be reported.
- The highest recorded velocity should be used to calculate PA pressure (unless beat-to-beat variability or respiratory variations exist).
- Pulmonic stenosis should always be ruled out by CW Doppler across the RVOT/pulmonic valve, documenting low (<1 m/sec) velocity; although pulmonic stenosis is rare in adult population, its absence should be verified in order to correctly assume that PA systolic pressure equals the RV systolic pressure.
- The peak-TR-velocity-based technique provides an estimation of pulmonary artery **systolic** pressure, hence cutoffs for presence and severity of pulmonary hypertension differ from those listed by the WHO (which are determined by mean PA pressure).

Pulmonary Pressure: Advanced

4

Abstract

Echocardiography can be utilized to assess pulmonary pressures. Pulmonary pressures are an important part of the hemodynamic assessment and carry significant prognostic and therapeutic implications. Most commonly, peak TR velocity is used to estimate pulmonary artery systolic pressure. While this measurement is commonly available and easily obtained, it provides only an estimate of the pulmonary artery systolic pressure, and not mean PA pressure. Mean PA pressure is the parameter used by the WHO to define presence and severity of pulmonary hypertension and carries important diagnostic and prognostic significance. In addition, evaluating pulmonary vascular resistance is critical for the proper classification of pulmonary hypertension and treatment considerations. Echocardiographic techniques are available to further assess pulmonary pressures including pulmonary artery diastolic pressure, mean pulmonary artery pressure, and pulmonary vascular resistance. Like with any other echo obtained measurement, attention to quality and accuracy of the interrogated flow and the obtained spectral signal is of paramount importance in order to assure the accuracy of the obtained data.

Keywords

Pulmonary hypertension · Pulmonary artery diastolic pressure · Mean pulmonary pressure · Pulmonary vascular resistance · Acceleration time

Introduction

As discussed in the previous chapter, echocardiography is frequently used to assess pulmonary artery pressure. Most commonly, peak TR velocity is used to estimate pulmonary artery systolic pressure. While this measurement is commonly available and easily obtained, it provides only an estimate of the pulmonary artery **systolic** pressure, and not mean PA pressure. Mean PA pressure is the parameter used by the WHO to define presence and severity of pulmonary hypertension and carries important diagnostic and prognostic significance. In addition, evaluating pulmonary vascular resistance is critical for proper classification of pulmonary hypertension and treatment considerations.

In this chapter, we will review available echocardiographic techniques to further assess pulmonary pressures. Using a case study, we will examine how to estimate the following parameters:
- Pulmonary artery diastolic pressure
- Mean pulmonary artery pressure
- Pulmonary vascular resistance

Like with any other echo obtained measurement, attention to quality and accuracy of the interrogated flow and the obtained spectral signal is of paramount importance; poor data in → unreliable data out.

4.1 Pulmonary Artery Diastolic Pressure

- Trace-to-mild pulmonic insufficiency (PI) is very common among healthy individuals.
- PI spectral Doppler (as recorded by CW Doppler) has several typical characteristics (Fig. 4.1).
  - Overall low-velocity signal.
  - "A dip" at end diastole

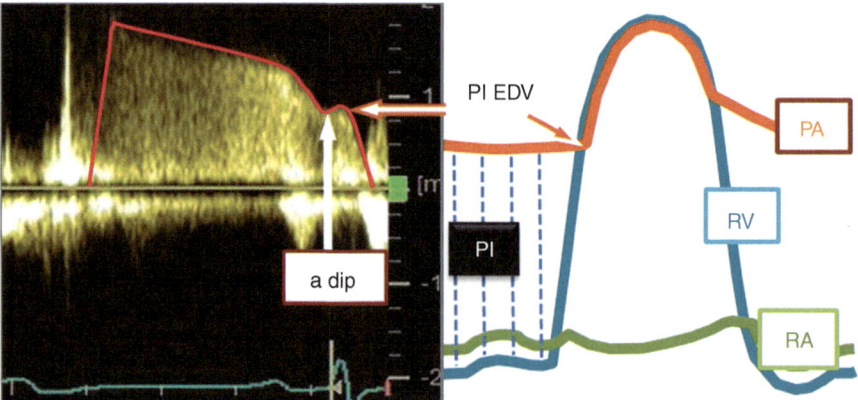

Fig. 4.1 Pulmonic insufficiency (PI) spectral Doppler. Continuous-wave Doppler tracing of PI typically shows an overall low-velocity signal with an "a dip" at end diastole (PA—pulmonary artery, RA—right atrium, RV—right ventricle)

- The driving pressure for PI is the diastolic pressure gradient between the pulmonary artery and the right ventricle (diastolic ΔP_{PA-RV}).
- Typically, this pressure gradient is low; normal PA diastolic pressure ~ 10 mmHg, normal RV diastolic pressure ~ 5 mmHg → diastolic ΔP_{PA-RV} up to 5 mmHg, which correlates with flow velocity of approximately 1 m/sec.
- At end diastole, atrial contraction occurs, which forces more blood into the RV.
- This slight increase in RV volume causes a slight increase in RV pressure, which **reduces** the pressure difference between the PA and the RV.
- Since the end-diastolic ΔP_{PA-RV} is reduced, the PI velocity decreases during atrial contraction.
- Notably, the "a dip" is not seen on aortic insufficiency (AI) spectral Doppler even though the same physiology applies.
  - The driving pressure for AI is the diastolic pressure gradient between the aorta and left ventricle (diastolic ΔP_{Ao-LV}).
  - Typically this pressure gradient is significantly larger as compared to the PI driving pressure gradient: normal aortic diastolic pressure ~ 70mmHg, normal LV diastolic pressure ~ 10 mmHg → diastolic ΔP_{Ao-LV} approximately 60 mmHg, which correlates with a flow velocity of 3.8 m/sec.
  - The slight volume/pressure increase in the LV during atrial contraction is negligible compared to the existing ΔP_{Ao-LV}, thus typically, no measurable change in AI velocity is seen on the AI spectral Doppler.
- Measuring the PI end-diastolic velocity allows calculation of the end-diastolic ΔP_{PA-RV}.
- During diastole, the tricuspid valve is open; RV diastolic pressure essentially equals RA pressure (assuming no tricuspid stenosis, which is rare).
- The PA diastolic pressure can thus be calculated as: Dias PAP = RAP + PI end diasΔP.
- Figure 4.2 shows tracings from a patient with severe pulmonary hypertension.
- Pulmonary artery systolic pressure, as estimated by the peak TR velocity is calculated to be 80 mHg.
  - Peak TR velocity 4 m/sec → ΔP_{RV-RA} = 64 mmHg
  - IVC dilated and plethoric → RAP estimated ~15 mmHg

$$\text{Sys PAP} = \text{RAP} + \text{TR}\Delta P = 64 + 15 \approx 80\,\text{mmHg}$$

- Pulmonary artery diastolic pressure is estimated based on the PI spectral Doppler.
- PI end-diastolic (ED) velocity ~ 2 m/sec → ΔP_{PA-RV} = 16 mmHg

$$\text{Dias PAP} = \text{RAP} + \text{ED PI}\Delta P = 15 + 16 \approx 30\,\text{mmHg}$$

4.2 Mean Pulmonary Artery Pressure

There are several techniques to estimate the mean pulmonary artery pressure.

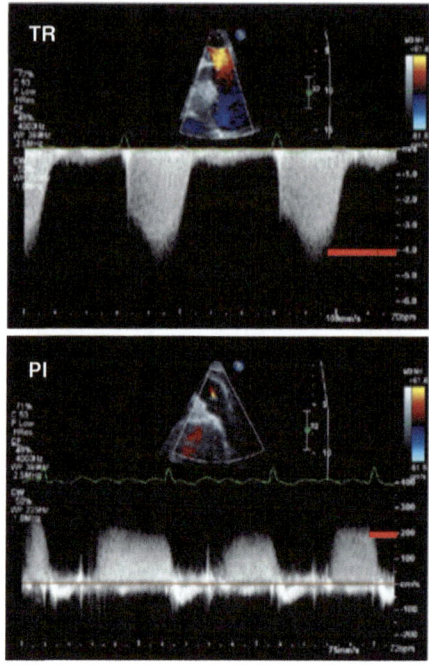
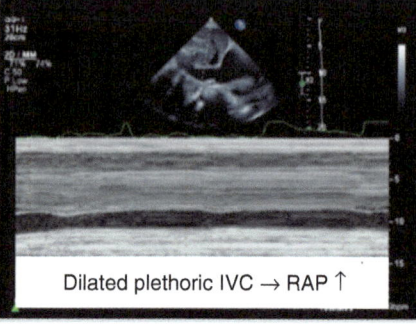

TR Peak V = 4m/sec, ΔP= 64mmHg

$PA_{sys}P = RV_{sys}P = \Delta P_{RV-RA} + RAP =$
64 + 15mmHg ≈ 80mmHg

Dilated plethoric IVC → RAP ↑

PI ED V = 2m/sec, ΔP= 16mmHg
$PA_{dia}P = \Delta P_{PA-RV} + RAP =$
16 + 15mmHg ≈ 30mmHg

Fig. 4.2 Severe pulmonary hypertension. Tricuspid regurgitation (TR) and pulmonic insufficiency (PI) tracings showing severely elevated systolic and diastolic pulmonary pressure

4.2.1 Calculating Mean PAP Based on Sys/Dias PAP

- Mean PAP can be calculated if PA systolic and diastolic pressures are known, using the formula for calculation of mean arterial pressure:

$$\text{mean PAP} = \frac{2}{3}\text{Dias PAP} + \frac{1}{3}\text{Sys PAP}.$$

- The formula is driven by the fact that diastole is significantly longer than systole; with a relatively normal heart rate, most of our life is spent in diastole.
- In the case study: Dias PAP ~ 30 mmHg, Sys PAP ~80 mmHg → mean PAP ~47 mmHg.

4.2.2 PI Early Diastolic Velocity

- Looking at the PI spectral Doppler, early diastolic velocity can also be measured (Fig. 4.3).
- There is a reasonable correlation between early diastolic PI pressure gradient and mean PA pressure.
- Mean PAP is calculated as: mean PAP = RAP + early PIΔP.
- In our case study, early PI velocity is approximately 2.5 m/sec.
- Thus, mean PAP = 15 + 4 × 2.5² =15 + 25 = 40 mmHg.

4.2 Mean Pulmonary Artery Pressure

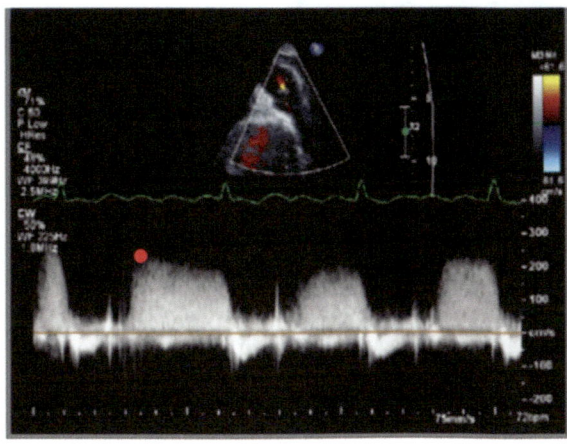

mean PAP = RAP + 4 X early PIv^2
PI Early$_v$ ~ 2.5m/sec →
mean PAP ~ 40mmHg

Fig. 4.3 Early diastolic pulmonic insufficiency (PI) velocity. The PI early diastolic velocity correlates reasonably well with mean pulmonary artery pressure

4.2.3 TR VTI

- The CW spectral Doppler envelope can be traced to calculate the TR VTI and the TR mean gradient (Fig. 4.4).
- A reasonable correlation exists between mean TR pressure gradient and mean PAP.
- Mean PAP is calculated as: mean PAP = RAP + mean TR gradient.
- In the case study: mean TR pressure gradient (by tracing the TR envelope) ~35 mmHg.
- Thus, mean PAP = 15 + 35 = 50 mmHg.

4.2.4 PA Acceleration Time

- The pulmonary outflow spectral Doppler envelope can also provide estimation of mean PAP.
- In general, the higher the pulmonary pressure (and pulmonary vascular resistance), the steeper the acceleration is (resembling aortic outflow spectral Doppler).
- The formulas used for calculating mean PAP based on pulmonary acceleration time (AT), perform best when heart rate is in the normal range (60–100 BPM).
- In general, mean PAP is calculated as: mean PAP = 79 − (0.45 × AT).
- If AT is <120 msec, mean PAP should be calculated as: mean PAP = 90 − (0.62 × AT).
- In the case study (Fig. 4.5): AT measured to be 79 mSec.
- Thus, mean PAP = 90 − (0.62 × 79) = 90 − 49 = 41 mmHg.

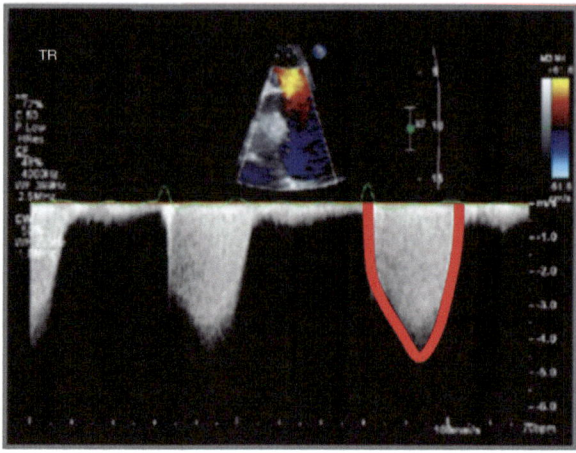

mean PAP = ΔP (RV − RA)$_{mean}$ + RAP
ΔP (RV-RA)$_{mean}$ ~ 35mmHg →
mean PAP ~ 50mmHg

Fig. 4.4 Tricuspid regurgitation (TR) envelope. Tracing the TR spectral envelope can give the mean RV–RA gradient ($\Delta P_{\text{(RV-RA) mean}}$) during systole. There is a reasonable correlation between $\Delta P_{\text{(RV-RA) mean}}$ and mean pulmonary artery pressure

4.3 Pulmonary Vascular Resistance

- Elevated pulmonary artery pressures can be the result of:
  - Increased flow (e.g., large intracardiac defect with significant left-to-right shunt causing a markedly elevated pulmonary blood flow).
  - Increased pulmonary vascular resistance (with normal pulmonary blood flow).
- Treatment and prognosis of these conditions can differ; distinction between them is imperative.
- Estimating pulmonary vascular resistance (PVR) can be of utmost importance, as mentioned, both for prognostic purposes as well as therapeutic purposes.
- It should be noted that echocardiography can provide a general ball park estimation of PVR; PVR should be further confirmed by invasive measurements, especially before therapeutic decisions are made (e.g., avoidance of ASD closure due to prohibitively elevated PVR).
- Normal PVR is approximately 1.5 Wood units (WU); for diagnosis of pulmonary arterial hypertension, the currently used cutoff is PVR > 3WU.
- There are two main ways to estimate PVR.

4.3.1 Comparing TR ΔP and RVOT Flow

- RVOT VTI can be used as a surrogate for pulmonary blood flow.
- The higher the pulmonary circulation blood flow, the higher the RVOT VTI.

4.3 Pulmonary Vascular Resistance

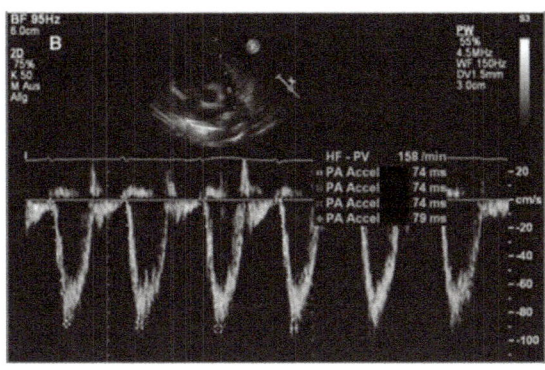

mean PAP = 79 − (0.45 × AT)
AT < 120m Sec → mean PAP = 90 − (0.62×AT)
Shorter AT → higher PVR → higher PAP

Fig. 4.5 Pulmonary outflow acceleration time (AT). AT can be used to estimate mean pulmonary pressure. In general, the higher the pulmonary pressure (and pulmonary vascular resistance), the steeper the acceleration with shorter AT

- Comparing the TR peak velocity to the RVOT VTI can provide a clue to whether the increased pulmonary pressure is due to high flow (in which case the RVOT VTI will be elevated) or high resistance (in which case the RVOT VTI will be relatively low).
- Higher RVOT VTI suggests lower PVR; lower RVOT VTI suggests higher PVR.
- Examining the ratio $\dfrac{\text{TR Peak Velocity}\left(\dfrac{m}{sec}\right)}{\text{RVOT VTI}(cm)}$ provides a ball park for normal vs. abnormal PVR.
- $\dfrac{\text{TR}}{\text{RVOT VTI}} > 0.12$ has excellent sensitivity for PVR > 1.5 WU.
- In our case study: TR peak velocity = 4 m/sec, RVOT VTI − 8.5 cm.
- $\dfrac{\text{TR}}{\text{RVOT VTI}}$ ratio is calculated to be $\dfrac{4}{8.5} = 0.47$, consistent with elevated PVR.
- PVR may also be estimated using the formula:

$$\text{PVR} = \dfrac{\text{TR Peak Velocity}}{\text{RVOT VTI}} \times 10 + 0.16.$$

- In our case study: $\text{PVR} = \dfrac{4}{8.5} \times 10 + 0.16 = 4.7 + 0.16 \approx 4.9\,\text{WU}$, suggestive of markedly elevated PVR.

4.3.2 Calculating Pressure Drop and Flow across Pulmonary Circulation

- Flow (F), resistance (R), and pressure difference (ΔP) are related by: $F = \dfrac{\Delta P}{R}$
- Rearranging the above formula: $R = \dfrac{\Delta P}{F}$.
- When calculating PVR: ΔP is the pressure drop across the pulmonary circulation, F is the pulmonary circulation cardiac output (CO).
    - The pressure drop across the pulmonary circulation is the difference between mean pulmonary artery pressure and left atrial pressure.
    - Mean PA pressure can frequently be estimated by echocardiography (see above).
    - Left atrial pressure can be often estimated by echocardiography utilizing mitral inflow spectral Doppler, tissue Doppler of the mitral annulus, left atrial size, and TR peak velocity; most commonly it can be estimated to be either "normal" or "elevated."
    - Cardiac output can be calculated using measurements of cross-sectional area and corresponding VTI; in the absence of intracardiac shunt, pulmonary CO equals systemic CO and hence any area where diameter and VTI can be obtained (e.g., LVOT, mitral annulus, and RVOT) may be utilized. However, in the presence of intracardiac shunt, care must be taken to use the correct CO: for PVR calculation, pulmonary CO must be calculated by measuring RVOT diameter and VTI.
- PVR is calculated as: $\text{PVR} = \dfrac{\text{mean PAP} - \text{LAP}}{\text{pulmonary CO}}$ (WU).
- In our case study: mean PAP estimated 40–45 mmHg, LAP was assessed as normal ~10 mmHg, and cardiac output as measured by LVOT diameter, VTI, and heart rate was estimated to be 6 lit/min.
- Thus, $\text{PVR} = \dfrac{40 - 10}{6} = 5$ WU.

4.4 Summary and Final Points

- Pulmonary artery systolic pressure is commonly estimated on echocardiography by utilizing peak TR velocity.
- A more detailed evaluation can be carried out by using other available data.
- Pulmonary artery diastolic pressure, mean pulmonary artery pressure, and pulmonary vascular resistance can frequently be estimated by echo.
- Confirming the calculated parameters by multiple techniques improves the accuracy of the evaluation.
- Notably, the parameters calculated by different techniques should fall in the general "ball park" of one another; it is not uncommon for different techniques to produce slightly different numbers.

4.4 Summary and Final Points

- Many assumptions underlie the various calculations used for the comprehensive evaluation of pulmonary hemodynamics; it is of paramount importance to be aware of possible pitfalls and measurement errors, in order to avoid misdiagnosis.
- If the signal quality is unreliable, spatial resolution for diameter measurement is suboptimal, etc. it is advisable to avoid reporting inconclusive data.
- Correlation with other findings on the echocardiogram is also important; RV size and function, right atrial size, presence and severity of TR and other valve lesions, pulmonic valve m-more, etc. The obtained data has to **make sense** and should **not** be evaluated in a vacuum.
- Lastly, important to remember—echocardiography cannot replace invasive assessment in complicated cases, especially when important prognostic and therapeutic decisions need to be made.

How Severe Is This MR

5

Abstract

Mitral regurgitation (MR) is one of the most common indications for echocardiography. Accurate quantification of MR severity is extremely important for proper decision-making regarding optimal treatment options. Echocardiography, and frequently trans-esophageal echo, provide thorough quantification information, as well as supporting evidence for the effects of the MR on the heart (e.g., LV/LA enlargement, presence and severity of pulmonary hypertension). It is important to utilize all available tools and measurements when assessing MR severity; reliance on one form of assessment can be deceiving and may result in inaccurate classification of MR severity. Conceptually, the severity of mitral regurgitation is determined by volume—how much blood (in milliliters) spills back from the LV into the LA during systole. A related concept that is also at the crux of determining MR severity is the regurgitant fraction—what percentage of the total stroke volume spills back into the LA. There are multiple echocardiographic quantification techniques for assessing MR severity including anatomic assessment for estimation of regurgitant orifice area, PISA method, color Doppler assessment, and volumetric calculations comparing flow volumes at various locations in the heart. For the most accurate assessment, a comprehensive evaluation should be undertaken and all possible data points should be analyzed.

Keywords

Mitral regurgitation · Regurgitant volume · Regurgitant fraction · Effective regurgitant orifice area (EROA) · Proximal iso-velocity surface area (PISA) · Volumetric calculations · Vena contracta · Direct planimetry

Introduction

Mitral regurgitation (MR) is one of the most common indications for echocardiography. Accurate quantification of MR severity is extremely important for proper decision-making regarding optimal treatment options. Echocardiography, and frequently trans-esophageal echo, provide thorough quantification information, as well as supporting evidence for the effects of the MR on the heart (e.g., LV/LA enlargement and presence and severity of pulmonary hypertension). It is important to utilize all available tools and measurements when assessing MR severity; reliance on one form of assessment can be deceiving and may result in inaccurate classification of MR severity.

In this chapter, we will use a case study to demonstrate multiple MR quantification techniques. We will review:
- General principles for assessing mitral regurgitation.
- Anatomic assessment for estimation of regurgitant orifice area.
- PISA method (Proximal Iso-velocity Surface Area).
- Color Doppler assessment.
- Volumetric calculations utilizing multiple data points to show how to verify accuracy of the various calculations.

5.1 General Principles

- Mitral regurgitation occurs when the mitral valve leaflets' coaptation is incomplete, allowing blood to spill back from the left ventricle into the left atrium during systole, when the mitral valve is supposed to be closed.
- Typically, MR is a holosystolic phenomenon, meaning it lasts throughout the entire mechanical systole—from the point of mitral valve closure, through isovolumetric contraction time, ejection time, and during isovolumetric relaxation time (thus covering the second heart sound), ending with mitral valve opening (Fig. 5.1).
- Certain pathologies result in less than holosystolic MR (e.g., mitral valve prolapse with late systolic MR); it is important to recognize these instances and use appropriate quantification techniques in order to avoid misinterpretation of the MR severity (see below).
- The driving pressure gradient of MR is high; it is the difference between systolic LV pressure (normally ~110–120 mmHg) and LA pressure (normally ~10 mmHg). Under normal hemodynamic conditions, the MR gradient is approximately 100 mmHg, corresponding to a peak MR velocity of 5 m/sec.
- It is important to understand that MR peak velocity is **not** a manifestation of the MR severity; it is a reflection of the pressure difference between the LV and LA during systole. It can be higher than usual or lower than usual under

Fig. 5.1 Holosystolic mitral regurgitation (MR). Typically, MR is a holosystolic phenomenon, meaning it lasts throughout the entire mechanical systole; from mitral valve closure (MVC), through isovolumetric contraction time, ejection time, and isovolumetric relaxation time (thus covering the second heart sound), ending with mitral valve opening (MVO)

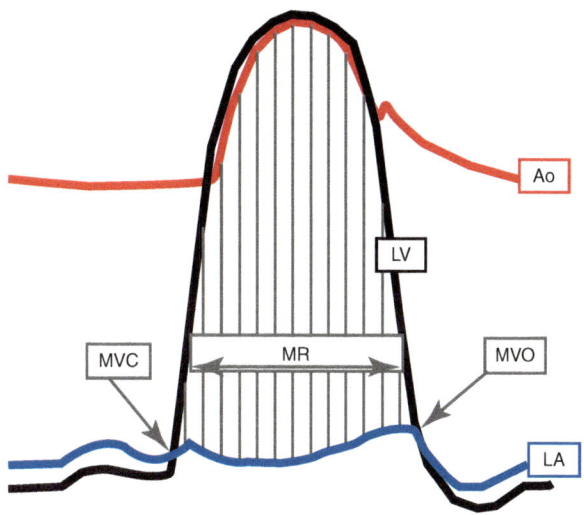

special conditions, however, this will be unrelated to the severity (i.e., quantity) of the MR.
- Conceptually, the severity of mitral regurgitation is determined by volume: how much blood (in milliliters) actually spills back from the LV into the LA during systole (regurgitant volume—RV).
- Related concept, which is also at the crux of determining MR severity, is the regurgitant fraction (RF): what percentage of the total stroke volume spills back into the LA.
- Whether regurgitant volume vs. regurgitant fraction is the ultimate determinant of MR severity remains unanswered; occasionally, when MR severity (by RV) is moderate, it can become important to calculate RF; if the RF is high, MR might need to be thought of as more severe than portrayed by the RV.
- When assessing the severity of mitral regurgitation, the aim is to assess the volume load on the left atrium (which ultimately results in elevated left atrial/pulmonary venous pressure, left atrial dilatation, left ventricular dilatation, etc.).
- Some quantification methods aim to calculate the regurgitant volume/fraction directly, whereas others aim to quantify surrogate parameters which relate to the regurgitant volume. For example, calculating the effective regurgitant orifice area (EROA) by any technique provides an assessment of the size of the orifice through which blood regurgitates back into the left atrium; naturally, the bigger the regurgitant orifice, the larger the regurgitant volume, hence the more severe the MR.
    - Yet, it is important to understand that effective regurgitant orifice area may not depict the true severity of the MR; in mitral valve prolapse with late systolic MR, the calculated EROA may be large, yet the overall severity of the MR may not be severe.

- Cutoffs for gradation of MR severity by EROA are based on the assumption that the regurgitant orifice is open (and hence blood is regurgitating back) throughout the entire systole.
- However, in late systolic MR, the regurgitant orifice may be large, yet open during only part of systole, allowing for a lower total volume to regurgitate back into the LA.
- Since MR severity is determined by the regurgitant volume, in instances with late systolic MR, EROA (by PISA or direct planimetry) should not be used to grade the MR severity. EROA may be used to calculate regurgitant volume (see below) and MR severity should be graded based on the RV.
- Similar considerations apply when looking at other quantification methods; for instance, color Doppler assessment should not rely on a single "freeze frame"; in addition to artifacts that may make the color area appear larger, one frame also does not take into account the duration of the MR and may lead to misclassification of MR severity.
• For the most accurate quantification of MR severity, as many techniques as possible should be utilized and all results should be taken into account when formulating the ultimate gradation of the MR severity.

5.2 Anatomic/Color Assessment for EROA

• As discussed above, effective regurgitant orifice area (EROA) is often used as a surrogate for MR severity.
• EROA can be measured/estimated in several ways.

5.2.1 3D-Based Direct Measurement of EROA

• Three dimensional (3D) echocardiography has advanced dramatically over the past 10–15 years, making it a readily available, clinically useful tool.
• Miniaturizing the 3D transducer such that it fits on a regular-sized TEE probe has revolutionized the clinical use of 3D TEE.
• There are several available imaging modalities utilizing 3D probe including full volume, full volume with color, 3D zoom, and narrow-angle acquisition.
• For direct measurement of mitral regurgitation EROA, 3D TEE should be used; the spatial resolution is significantly better as compared to TTE and imaging from the left atrial perspective allows for proper visualization of the EROA.
• EROA should be measured by post-processing of a full volume color image (Fig. 5.2).
• The processing software allows moving the 3-orthogonal planes, such that an en-face view of the EROA can be obtained. Note that a systolic frame must be chosen (generally prior to manipulation of the imaging planes) in order to obtain an image of the turbulent MR jet (rather than diastolic mitral inflow).

5.2 Anatomic/Color Assessment for EROA

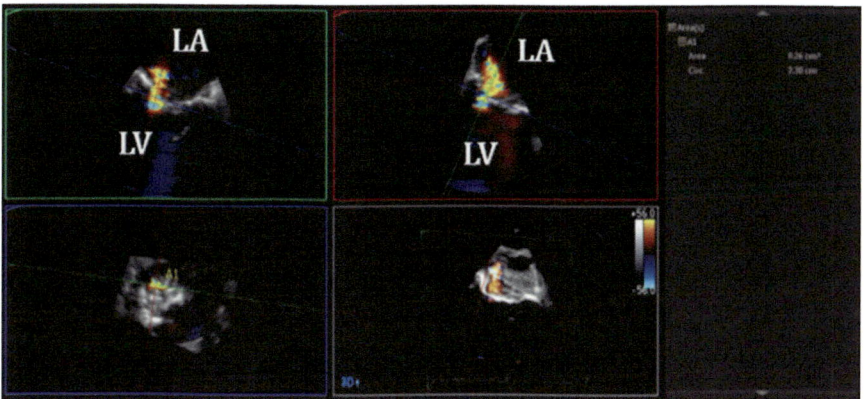

Fig. 5.2 Direct planimetry of effective regurgitant orifice area (EROA). For accurate measurement, full-volume color image should be acquired by 3D TEE. The processing software allows moving the 3-orthogonal planes, such that an en-face view of the EROA can be obtained. Once the imaging planes are oriented appropriately, direct planimetry of the EROA is performed by tracing the area of the turbulent jet, at the valve closure level (LA—left atrium, LV—left ventricle)

- Once the imaging planes have been oriented appropriately, direct planimetry of the EROA is performed by tracing the area of the turbulent jet, at the valve closure level.
- In our case study, the directly measured EROA came to 0.26 cm^2, which is in the moderate range.

5.2.2 Vena Contracta Measurement

- Vena contracta (VC) is the narrowest point of the MR jet.
- Typically, it is measured on the atrial side of the MR jet, immediately proximal to the valve closure point (Fig. 5.3).
- Vena contracta is a surrogate measurement for the EROA; it does not equal the EROA; VC is one dimensional, measured in length units (e.g., cm), EROA is two-dimensional, measured in area units (e.g., cm^2).
- In order to obtain an accurate vena contracta, the MR jet needs to be clearly visualized both distal and proximal to the valve closure point.
- Vena contracta must be measured on an appropriate view; in general, two-chamber view (or the similar mid esophageal, inter-commissural view on TEE) is not recommended for VC measurement. In these views, the mitral valve coaptation is visualized in cross section, perpendicular to the view of the VC.
- In instances where there is more than one MR jet, VC measurement may not serve as a good surrogate for the EROA since it might represent only one orifice through which MR occurs, and may not provide the whole picture.
- In our case study, the VC measures 0.32 cm, which is in the moderate range.

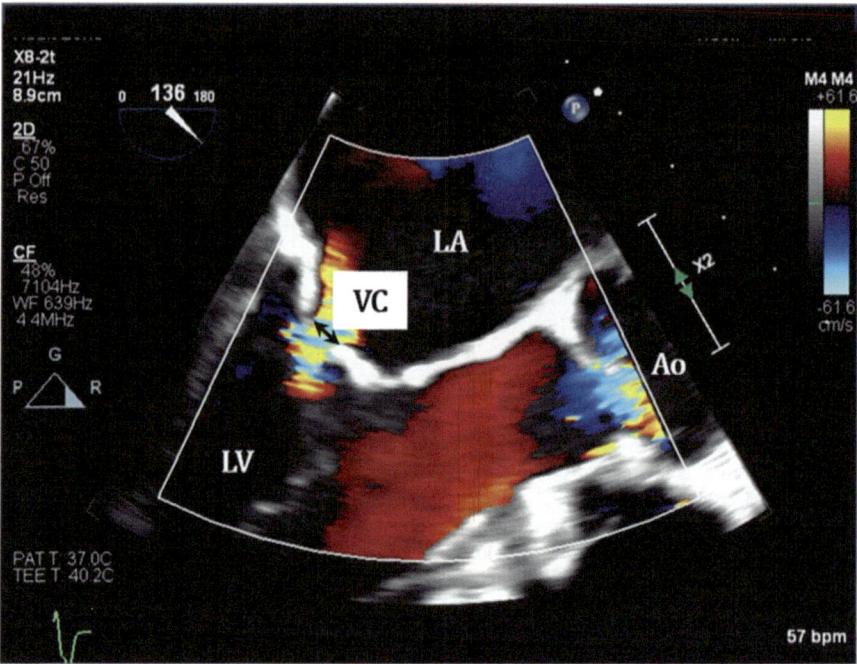

Fig. 5.3 Vena contracta (VC). Systolic frame of LVOT views from TEE. Typically, VC is measured on the atrial side of the MR jet, immediately proximal to the valve closure; it is a one-dimensional measurement, measured in length units (e.g., cm) (LA—left atrium, LV—left ventricle)

5.3 PISA Method

- The PISA method is an important quantification method by which to obtain both EROA and regurgitant volume (see also Chap. 2).
- PISA method is based on the continuity principle, comparing flow proximal to the EROA and the flow through the EROA.
- Importantly, since the method is based on comparing volumes through one valve, it can be used in the presence of other valve lesions or shunts.
- As the mitral regurgitation jet approaches the EROA, the flow converges toward the EROA.
- The jet converges to form smaller and smaller hemispheres, until it passes through the smallest orifice on its path—the regurgitant orifice.
- As the flow converges toward the regurgitant orifice, flow velocity accelerates; by the continuity principle, if the same volume of flow must pass through a smaller area at the same time → velocity must increase.

5.3 PISA Method

- When the flow converges toward the regurgitant orifice, it creates concentric hemispheres of smaller and smaller diameters; for each hemisphere the velocity at any point on the surface of the hemisphere is equal.
- Color Doppler allows identification of the hemisphere where the surface velocity is **exactly** the Nyquist limit.
    - In general, color Doppler is based on average jet velocity; by convention, flow toward the transducer is coded in red, and flow away from the transducer is coded in blue.
    - Notably, color Doppler imaging does not allow for accurate jet velocity calculation.
    - When the jet's average velocity passes the Nyquist limit (which is determined by the depth of imaging), the color changes from a laminar red or blue to a turbulent yellowish color.
    - Knowing the Nyquist limit (which is shown on the color scale of a color Doppler image) allows knowing the exact velocity at the point where the color signal changes from laminar to turbulent.
- The convergence hemisphere where the color signal changes from laminar to turbulent signal can be identified on the color Doppler image.
- At the surface of this identifiable hemisphere, the jet velocity equals the Nyquist limit.
- Knowing the surface area of the hemisphere and the velocity at the surface allows calculation of the flow rate at that area.
    - Surface area (SA) of hemisphere is given by the formula: $SA = 2\pi r^2$ (where r is the measured radius of the hemisphere).
    - Thus, the flow rate at the PISA surface is: $2\pi r^2 \times V_{Nyquist}$.
- The highest velocity of the jet (MR V_{max}) occurs **at** the regurgitant orifice area, since this is the smallest orifice along the jet's path.
    - Thus, the flow rate at the regurgitant orifice is: EROA × MR V_{max}.
- By the continuity principle, the flow rate at the PISA surface **equals** the flow rate at EROA.
- Meaning: $2\pi r^2 \times V_{Nyquist}$ = EROA × MR V_{max}.
- Rearranging the formula allows solving for EROA: $EROA = \dfrac{2\pi r^2 \times V_{Nyquist}}{MR\ V_{max}}$.
- Similar to other volume calculations (see Chap. 2), once the mitral regurgitation EROA is known, flow volume through that orifice (the regurgitant volume, RV) can be calculated by using the MR VTI.
- The regurgitant volume is: RV = EROA × MR VTI.
- In our case study (Fig. 5.4): PISA r = 0.7 cm, $V_{Nyquist}$ = 38.5 cm/sec, MR V_{max} = 557 cm/sec, MR VTI = 196 cm.
- Using the above parameters:

$$EROA = \frac{2 \times 3.14 \times 0.7 \times 0.7 \times 39}{557} = 0.22\,cm$$

-
- $RV = 0.22 \times 196 = 43\,cc$

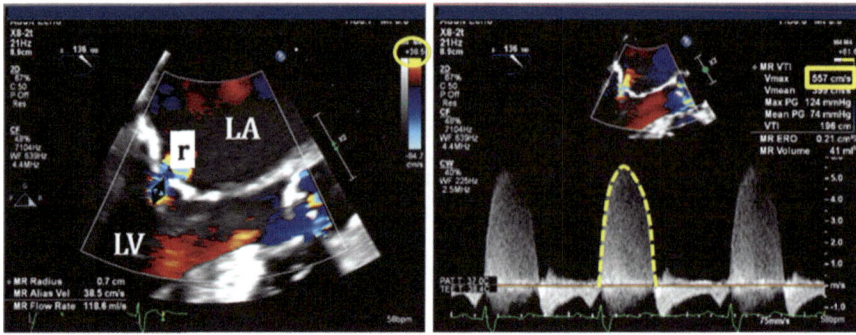

$PISA_r = 0.7\text{cm}$, $V_{nyquist} = 38.5\text{cm/sec}$, MR $V_{max} = 557\text{cm/sec}$

$2\pi PISAr^2 \times V_{nyquist} = EROA \times MR\ V_{max}$

$EROA = 0.22\text{cm}^2$, $RV = 43\text{cc}$

Fig. 5.4 Proximal iso-velocity surface area (PISA) method. The PISA method is based on the continuity principle, comparing flow proximal to the EROA and flow through the EROA. As the flow converges toward the regurgitant orifice, concentric hemispheres of smaller and smaller diameters are formed, and flow velocity accelerates. The convergence hemisphere where the color signal changes from laminar to turbulent signal can be identified and its radius (*r*) measured. The Nyquist limit is shown on the color Doppler image (*yellow circle* around the color bar), allowing calculation of the flow rate at the PISA surface. Tracing the MR spectral envelope provides the MR VTI, which can be used, along with the calculated EROA, to estimate the regurgitant volume (RV) (LA—left atrium, LV—left ventricle)

- Both these values are in the moderate range; notably, the EROA is very similar to that obtained by direct planimetry using 3D post-processing (see above), further confirming the accuracy of the measurement.

5.3.1 Pointers and Potential Pitfalls for the PISA Method

It is important to pay meticulous attention to technical details when measuring the PISA in order to obtain an accurate measurement of the EROA and RV.

- Imaging view should be optimized to obtain the clearest PISA shell.
- Nyquist limit: The Nyquist limit should be changed such that the PISA shell that is obtained is larger, making measurement easier and more precise.
  - This can be achieved by either lowering the Nyquist limit, or moving the color baseline in the direction of the MR flow (note that this means moving the baseline down if imaging by TTE and up if imaging by TEE).
  - It is important to understand that using a lower Nyquist limit (by either one of these techniques) is meant only to improve the ability to accurately measure the PISA radius, which is generally only several millimeters in length. By

using a lower Nyquist limit, the PISA shell (which is the concentric hemisphere where the jet velocity equals the Nyquist limit) will be larger, making a precise measurement of the radius more likely.
    - However, it should obviously not affect the calculated flow rate: if a higher Nyquist limit is used, the PISA shell (and hence r) will be smaller, such that the product $2\pi r^2 \times V_{Nyquist}$ will remain the same as if a lower Nyquist limit was used (and a larger r measured).
- Eccentric jets or multiple jets may not form concentric hemispheric shells; in these cases, PISA may not be the preferred way to quantify MR.
- The PISA radius should be measured from the PISA shell to the mitral valve closure line; occasionally a "color compare" image is needed in order to identify the closure point of the mitral leaflets.

5.4 Color Doppler Assessment

- The most common way to estimate MR severity is by looking at the color jet.
- While this technique is easy and readily available, there are potential pitfalls to be aware of and take into consideration.
- In general, when looking at a color Doppler signal several aspects of the jet are assessed:
    - Absolute jet size.
    - Jet size relative to left atrial area.
    - Jet width (see above—vena contracta).
    - Presence of coanda effect.
- These color jet parameters are affected by MR severity, however, by other factors as well.
    - Color gain settings, Nyquist limit, wall filters—all can change the appearance of a color jet and hence should be optimized for appropriate MR imaging.
    - Left atrial size may complicate jet size estimation; normal left atrial size may cause a jet to appear as if it takes a larger percentage of the LA area. Conversely, dilated LA may cause the MR jet to occupy a smaller percentage of its area.
    - Different views may be more or less suitable for any particular MR jet; it is important to image from multiple views in order to obtain a comprehensive evaluation of the MR jet.
    - Eccentric jets may be more prone to the coanda effect, where the jet impinges upon the left atrial wall, making it appear smaller and hence causing possible underestimation of the MR severity.
- As mentioned above, the duration of the mitral regurgitation may have a significant impact on the overall severity of the MR; shorter jets (e.g., late systolic MR in mitral valve prolapse) may be less severe than appears if only a single color Doppler frame is analyzed for jet size and relative area.
- Looking at our case study (Fig. 5.5) the color jet appears to be less than 10 cm^2 and occupies approximately a third of the LA size; these parameters are in the moderate range.

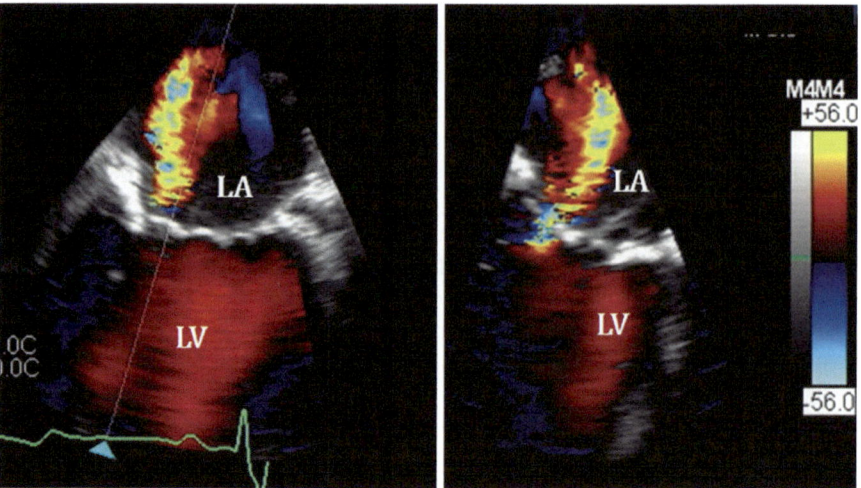

Fig. 5.5 Color Doppler assessment of MR. Systolic frame of biplane view from TEE showing turbulent MR jet in the left atrium (LA). The jet appears to occupy approximately a third of the LA area (LV—left ventricle)

5.5 Volumetric Calculations

- The underlying principle for volumetric calculations is the conservation of mass; every drop of blood needs to be accounted for.
- Comparing flow volumes between different valves should allow us to calculate any "extra" volume anywhere.
- However, it should be noted that if more than single valve regurgitation is present or if there is an intracardiac shunt, volumetric calculations may not be adequate to resolve the different volumes.
- Typically, the inflow into the LV (through the mitral valve) is compared with the outflow out of the LV (through the LVOT) (Fig. 5.6).
- Under normal conditions, with no valve regurgitation, these volumes are equal.
- However, if mitral regurgitation is present, the inflow into the LV includes the volume that spilled back into the LA during the preceding systole (the RV), whereas the outflow through the LVOT does not (rather the LVOT flow represents the "effective" stroke volume).
- The inflow through the mitral valve is calculated by measuring the mitral annular area and the pulse-Doppler VTI at the annulus (note, this VTI is different than the routinely obtained PW at the tips of the mitral valve).
- The outflow across the LVOT is calculated by measuring the LVOT CSA and the LVOT VTI.

5.5 Volumetric Calculations

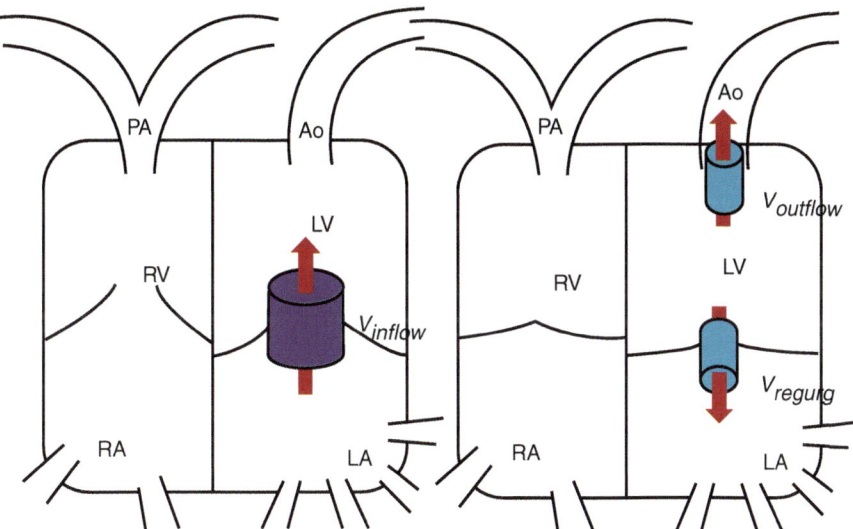

Fig. 5.6 Volumetric calculation of mitral Regurgitant Volume (RV). The mitral inflow volume (V_{inflow}) can be calculated by measuring the mitral annulus diameter and VTI. The outflow volume from the left ventricle ($V_{outflow}$) can be calculated by measuring the LVOT diameter and VTI. The difference between these two volumes is the RV (Ao—aorta, LA—left atrium, LV—left ventricle, PA—pulmonary artery, RA—right atrium, RV—right ventricle)

- From these two calculations, the RV can be obtained:
- RV = MV annular inflow − LVOT outflow.
- RV = $\pi \times r_{annulus}^2 \times MV_{annulus} VTI - \pi r_{LVOT}^2 \times LVOT\ VTI$.
- Alternatively, the mitral annulus CSA and LVOT CSA can be measured directly by post-processing 3D images; this allows for fewer geometric assumptions as well as avoidance of mistakes that are caused by squaring an inaccurately measured diameter.
- In our case study (Figs. 5.7 and 5.8): LVOT CSA (directly measured) = 3.2 cm^2, LVOT VTI = 24 cm, MV annular area (directly measured) = 5 cm^2, MV annular VTI = 23 cm.
- Using the above numbers:
  MV inflow = 5 × 23 = 115cc, LVOT outflow = 3.2 × 24 ≈ 80cc.
  RV = MV inflow − LVOT outflow = 115 − 80 = 35cc.
- Once again, this RV is in the moderate range, and very similar to the RV obtained by the PISA method.
- In this case study, the RVOT was also evaluated; RVOT diameter = 2.9 cm, and RVOT VTI = 13 cm (Fig. 5.9).

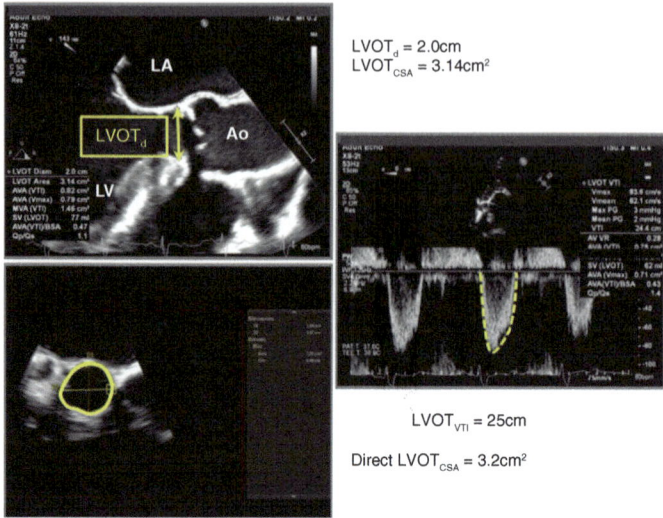

Fig. 5.7 Calculating outflow through the LVOT. TEE images from MR quantification case study. LVOT cross-sectional area (LVOT$_{CSA}$) calculated from LVOT diameter (LVOT$_d$) measurement as well as by post-processing and direct planimetry of a 3D image. Stroke volume calculated utilizing the LVOT$_{CSA}$ and LVOT$_{VTI}$ (Ao—aorta, LA—left atrium, LV—left ventricle)

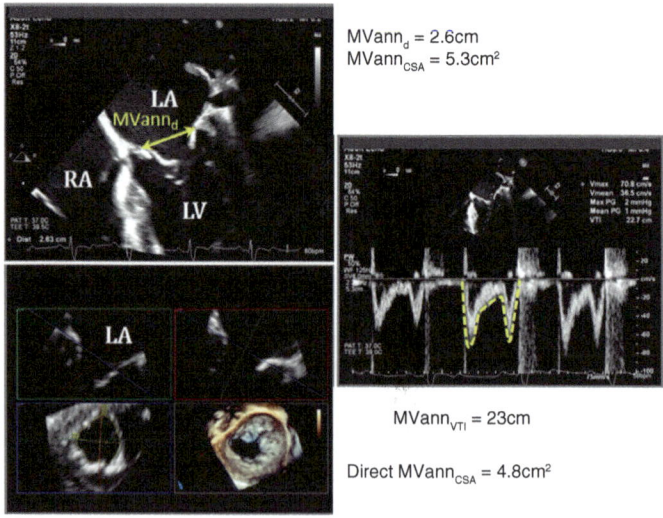

Fig. 5.8 Calculating inflow through the Mitral Valve Annulus. TEE images from MR quantification case study. Mitral annular cross-sectional area (MVann$_{CSA}$) calculated from annular diameter (MAann$_d$) measurement as well as by post-processing and direct planimetry of a 3D image. Stroke volume calculated utilizing the MVann$_{CSA}$ and MVann$_{VTI}$ (LA—left atrium, LV—left ventricle, RA—right atrium)

5.5 Volumetric Calculations

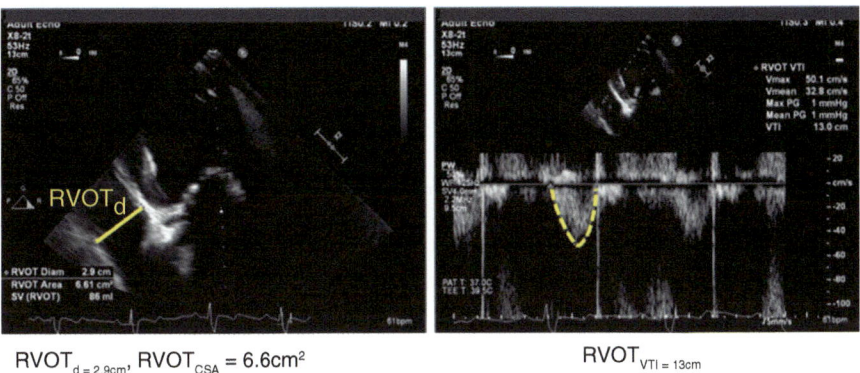

$RVOT_{d = 2.9cm}$, $RVOT_{CSA} = 6.6 cm^2$ $RVOT_{VTI = 13cm}$

SV_{right} = CSA × VTI = 6.6 × 13 = 85cc

Fig. 5.9 Calculating outflow through the RVOT. TEE images from MR quantification case study. RVOT cross-sectional area ($RVOT_{CSA}$) calculated from RVOT diameter ($RVOT_d$) measurement. Stroke volume calculated utilizing the $RVOT_{CSA}$ and $RVOT_{VTI}$

- Using these numbers, RVOT outflow was calculated: RVOT outflow = RVOT CSA × RVOT VTI = πr_{RVOT}^2 × RVOT VTI = 3.14 × 1.45 × 1.45 × 13 = 85 cc.
- The calculated RVOT outflow was very similar to the calculated LVOT outflow, further confirming the accuracy of the above calculations.

5.6 Summary and Final Points

- When assessing MR severity, a comprehensive evaluation should be undertaken and all possible data points should be analyzed.
- Anatomy and underlying pathology causing the mitral regurgitation should be assessed; extreme cases can disclose the severity of the MR (i.e., very severe) even before color Doppler has been looked at (for instance when there is a very large flail gap or severely torn and infected mitral leaflet).
- In addition to the techniques described above there are more quantification methods (qualitative and quantitative) that may be used (e.g., pulmonary vein flow pattern and measuring peak mitral inflow velocity).
- Assessing for supportive signs that indicate the MR effect on the heart is essential for obtaining the full picture of the MR severity; these may include, LV/LA volume, LV ejection fraction, presence and severity of pulmonary hypertension, and more.
- Although in the majority of cases, echocardiography (TTE and TEE) is sufficient for detailed assessment of MR cause and severity, in cases where there are conflicting data, other imaging modalities may be used in addition to echocardiography; these may include cardiac MRI or invasive LV angiography.

What Is Wrong with This MR: Part I

Abstract

Mitral regurgitation (MR) is a common valve disease, frequently encountered in echocardiography laboratories. Correct quantification of the severity of the MR is of paramount importance for accurate prognostication and proper treatment planning. The MR spectral Doppler tracing can provide important information regarding intracardiac pressures and be a clue to the presence of various cardiovascular pathologies. Assessing the MR jet velocity is essentially unrelated to the MR severity. Severity is determined by the quantity of the MR—the volume (or fraction) of blood that spills back into the LA instead of being ejected forward through the aortic valve. MR velocity is determined by the pressure gradient driving this velocity, irrespective of the volume of blood that regurgitates back into the LA. Analyzing MR spectral Doppler can provide a clue to the presence of various cardiac (or extra-cardiac) pathologies, and presence of an unusual MR spectral tracing should prompt a search for its cause.

Keywords

Mitral regurgitation · High-velocity MR · Subclavian stenosis

Introduction

- Mitral regurgitation is a common valve disease encountered frequently in echocardiography.
- Correct quantification of the severity of the MR is of paramount importance for accurate prognostication and proper treatment planning.
- The MR spectral Doppler tracing can provide important information regarding intracardiac pressures and be a clue to the presence of various cardiovascular pathologies.

- It is imperative to understand that assessing the MR jet velocity is essentially **unrelated** to the MR severity.
    - As discussed in the previous chapter, severity is determined by the quantity of the MR—the volume (or fraction) of blood that spills back into the LA instead of being ejected forward through the aortic valve.
    - MR velocity is determined by the pressure gradient driving this velocity, irrespective of the volume of blood that regurgitates back into the LA.
- In this chapter, we will review a case study that demonstrates how to interpret unusual MR velocity.

6.1 Case Presentation

- A 63-year-old woman was referred for echocardiogram to assess her left ventricular ejection fraction due to symptoms of dyspnea on exertion.
- Left ventricular function was mildly reduced, with segmental wall motion abnormalities; hypokinesis of the inferior septum, basal inferior wall, and inferolateral wall were noted.
- Moderate mitral regurgitation was seen, secondary to mitral valve tethering (full quantification not shown) (Fig. 6.1).

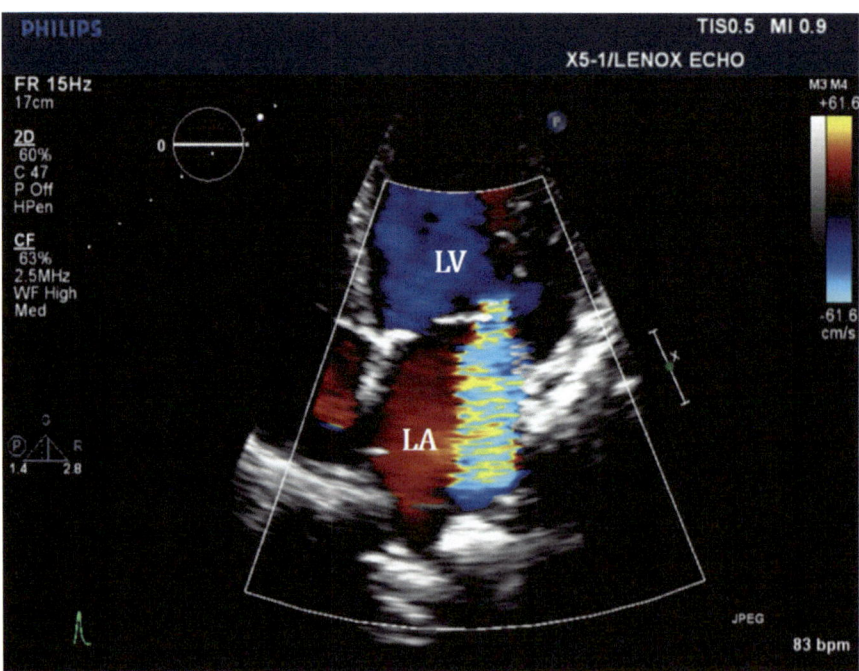

Fig. 6.1 Mitral regurgitation (MR). Systolic frame from apical four-chamber view showing turbulent MR jet. The jet occupies approximately 40% of the left atrial area. Other quantification parameters not shown (LA—left atrium, LV—left ventricle)

6.2 Approach to an Unusual MR Tracing

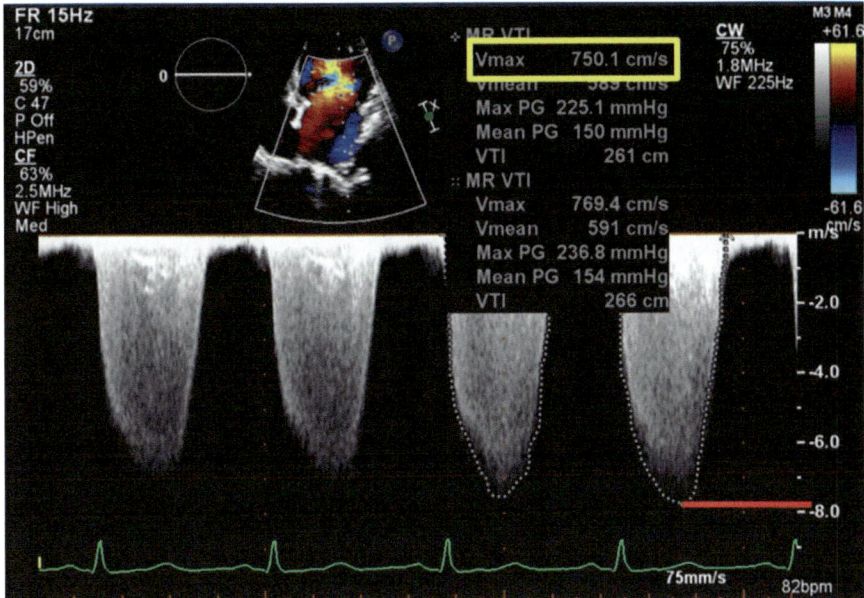

Fig. 6.2 Mitral regurgitation (MR) spectral Doppler. Continuous-wave Doppler of the MR from the presented patient. The MR envelope is unusual due to the very high peak velocity—approximately 7.5 m/sec

- The spectral Doppler of the MR jet was unusual (Fig. 6.2); the peak MR velocity was approximately 7.5 m/sec.

6.2 Approach to an Unusual MR Tracing

- Using the simplified Bernoulli equation ($\Delta P = 4v^2$), MR peak velocity of 7.5 m/sec means a driving pressure gradient of $4 \times 7.5^2 \approx 225$ mmHg.
- Whenever a pressure gradient is estimated, the next two questions are:
    - Between what two chambers this pressure gradient occurs.
    - At what part of the cardiac cycle.
- Regarding MR: The calculated ΔP represents the pressure difference between left ventricular pressure and left atrial pressure during systole (ΔP_{LV-LA}).
- Thinking about "normal" MR gradient (Fig. 6.3).
    - Left ventricular systolic pressure is generally ~110–120 mmHg (LV systolic pressure typically equals the aortic systolic pressure).
    - Normal left atrial pressure is generally ~10 mmHg.
    - Pressure difference between the LV and LA during systole: $\Delta P_{LV-LA} = 110 - 10 = 100$ mmHg.
    - ΔP_{LV-LA} of 100 mmHg corresponds to peak velocity of 5 m/sec.
    - This means, that under normal conditions, MR jet velocity is generally approximately 5 m/sec.

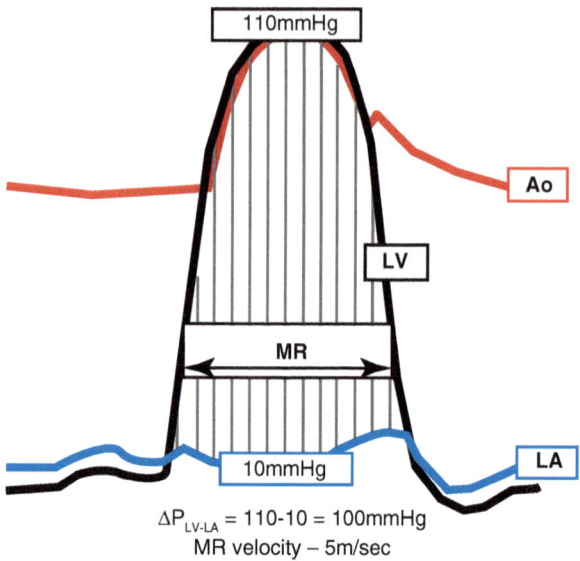

Fig. 6.3 Typical mitral regurgitation (MR) driving pressures. MR velocity is determined by the systolic pressure gradient between the left ventricle (LV) and the left atrium (LA)—ΔP_{LV-LA}. Under "normal" circumstances, ΔP_{LV-LA} equals approximately 100 mmHg, corresponding to a peak velocity of 5 m/sec (Ao-aorta)

- In the presented case, the MR velocity is significantly higher, demanding a search for the cause of this unusual velocity.
- The calculated ΔP of 225 mmHg means that the LV systolic pressure reaches a value that is 225 mmHg higher than LA pressure; if normal LA pressure is assumed (~10 mmHg), it means that peak LV systolic pressure reaches 235 mmHg.
- Thinking about this dramatically elevated LV systolic pressure of 235 mmHg (Fig. 6.4):
  - LV systolic pressure may be elevated and **equal** to aortic pressure.
  - LV systolic pressure may be elevated and **not equal** to aortic pressure.
- Obviously, the simplest way to distinguish between these two options is to measure the aortic pressure by taking the patient's blood pressure.
  - The patient's blood pressure was measured (on the left arm) at the time of the echo and found to be 155/88 mmHg.
  - While this blood pressure is elevated, it is not elevated enough to explain a systolic LV pressure of 235 mmHg.
- Meaning, the LV systolic pressure is elevated and **not equal** to the aortic pressure.
- In order for LV systolic pressure to be higher than aortic systolic pressure, there must be a pressure gradient somewhere between the LV and measured aortic pressure.
- The magnitude of this pressure gradient can be estimated:
- LV systolic pressure 235 mmHg, aortic systolic pressure 155 mmHg → $\Delta P_{LV-Ao} = 80$ mmHg.
- To reiterate, at this point it can be deduced that somewhere between the LV and the measured aortic pressure, there is a significant obstruction, causing a pressure gradient of approximately 80 mmHg.

6.2 Approach to an Unusual MR Tracing

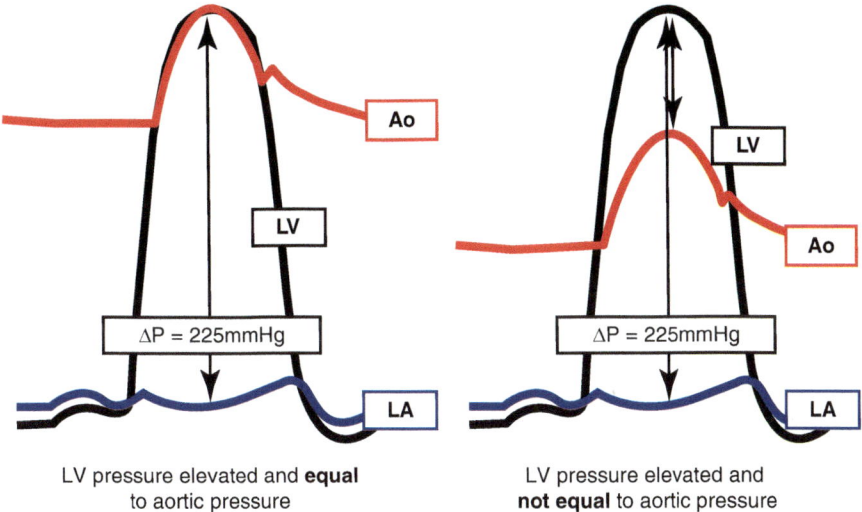

LV pressure elevated and **equal** to aortic pressure

LV pressure elevated and **not equal** to aortic pressure

Fig. 6.4 Two possible scenarios for high-velocity mitral regurgitation (MR). High-velocity MR implies high left ventricular (LV) systolic pressure. LV systolic pressure may be elevated and equal to aortic (Ao) pressure (*left diagram*), or they may be a gradient between the LV and the site where Ao pressure is measured (*right diagram*)

- The obstruction can be **anywhere** along the way between the LV and the site where the blood pressure was measured; it may be intra- or extra-cardiac; the mere presence of the elevated LV systolic pressure does not disclose where the obstruction is; a further investigation must be performed to find the site of obstruction.
- Possible locations of obstruction to flow between the LV and the measured blood pressure include:
  - LVOT.
  - Aortic valve.
  - Aorta—ascending, arch, proximal descending.
  - Left subclavian artery.
- All these sites are in between the LV and the site of blood pressure measurement and should be interrogated appropriately.
- Pulse Doppler at the LVOT and continuous wave Doppler across the aortic valve did not demonstrate any evidence for LVOT obstruction, aortic stenosis, or supravalvular aortic stenosis at the ascending aorta (Fig. 6.5).
- Doppler across the proximal descending aorta showed normal flow velocity, with no evidence for aortic coarctation.
- Limited Doppler study of the left subclavian artery was performed, showing increased velocity (up to 3.6 m/sec) confirming the presence of left subclavian stenosis (Fig. 6.6).

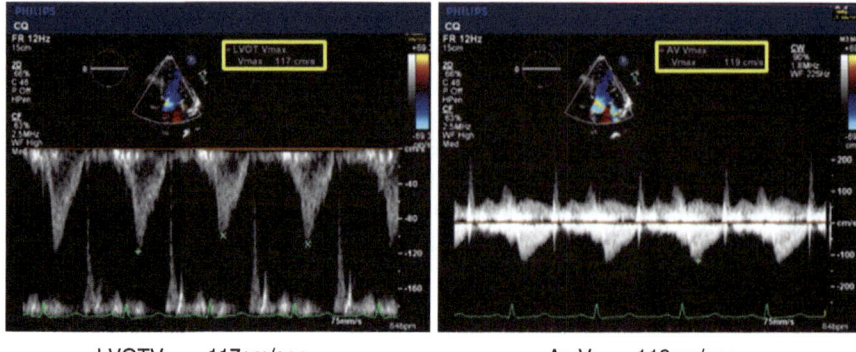

LVOTV$_{max}$ = 117cm/sec   Ao V$_{max}$ = 119cm/sec

Fig. 6.5 LVOT and aortic valve spectral Doppler. In order to verify there is no intra-cardiac gradient between the left ventricle and the aorta, pulse wave Doppler at the LVOT (*left panel*) and continuous wave Doppler through the LVOT-aortic valve tract (*right panel*) were obtained. No flow acceleration was noted in either one

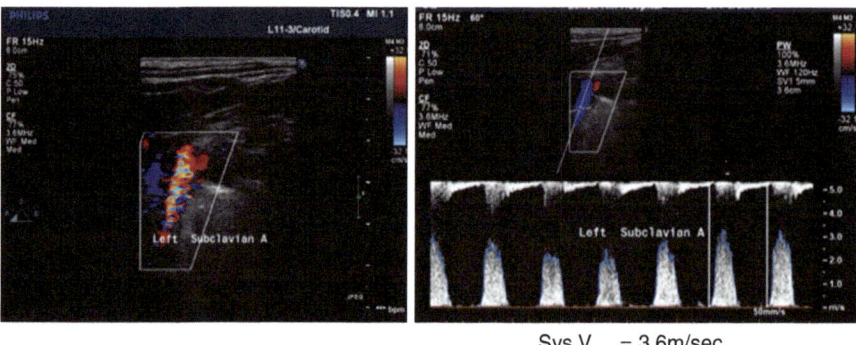

Sys V$_{max}$ = 3.6m/sec

Fig. 6.6 Left subclavian artery Doppler. Color Doppler (*left panel*) and spectral Doppler at the left subclavian artery showing turbulent flow with a peak velocity of 3.6 m/sec. This finding confirms the presence of a gradient between the left ventricle and the site where blood pressure was measured

6.3 Summary and Final Points

- In this case study, unusually high MR velocity was detected by the spectral Doppler.
- This finding did not imply anything relating to MR severity, however, disclosed a highly elevated LV systolic pressure.
- The measured blood pressure at the time of the echo was lower than the estimated LV systolic pressure, suggesting the presence of an obstruction with a significant gradient between the LV and the blood pressure measurement site.

6.3 Summary and Final Points

- It is important to remember that an obstruction between the LV and the site of blood pressure measurement may be intra-cardiac (e.g., LVOT obstruction/aortic stenosis) or extra-cardiac (e.g., aortic coarctation, subclavian stenosis).
- It should be noted, that another simple way to have made this diagnosis would be to measure the patient's blood pressure on her right arm; the expected right arm blood pressure should have been significantly higher than that measured on the left. Unfortunately, this could not have been done as the patient was post right mastectomy with lymph node dissection and was advised against any right arm blood pressure measurements.
- It is important to remember that when indicated, blood pressure should be measured on all four extremities; this may aid in understanding unusual flow patterns.

What's Wrong with This MR: Part II

Abstract

Mitral regurgitation (MR) is a common valve disease encountered frequently in echocardiography laboratories; correct quantification of the severity of the MR is of paramount importance for accurate prognostication and proper treatment planning. The MR spectral Doppler tracing can provide important information regarding intracardiac pressures and be a clue to the presence of various cardiovascular pathologies. MR jet velocity is essentially unrelated to the MR severity. Severity is determined by the quantity of the MR—the volume (or fraction) of blood that spills back into the LA; MR velocity is determined by the pressure gradient driving this velocity, irrespective of the volume of blood that regurgitates back into the LA. Unusual MR spectral tracing can manifest as abnormally high or abnormally low peak velocity, as well as slower rate of rise of the MR velocity. Identifying an unusual MR spectral tracing should prompt a search for its cause.

Keywords

Low-velocity MR · Systolic dysfunction · Cardiogenic shock · dP/dT

Introduction

- In the previous chapter, we saw that the MR spectral Doppler tracing can provide important information regarding intra-cardiac pressures and can be a clue to the presence of various cardiovascular pathologies.
- As mentioned before, assessing the MR jet velocity is essentially **unrelated** to the MR severity.
  - Velocity is determined by the pressure gradient driving the flow, irrespective of the flow volume.

- Severity is determined by the volume (or fraction) of blood that spills back into the LA rather than flowing forward through the aortic valve.
- In this chapter, we will review another case study that demonstrates how to interpret unusual MR velocity.

7.1 Case Presentation

- A 71-year-old man was admitted for shortness of breath and fatigue.
- Echocardiogram showed dilated left ventricle, diffuse hypokinesis, and overall severely reduced LV function.
- Mild-to-moderate mitral regurgitation was seen, secondary to annular dilatation.
- The spectral Doppler of the MR jet was unusual (Fig. 7.1); low peak velocity and irregular shape.

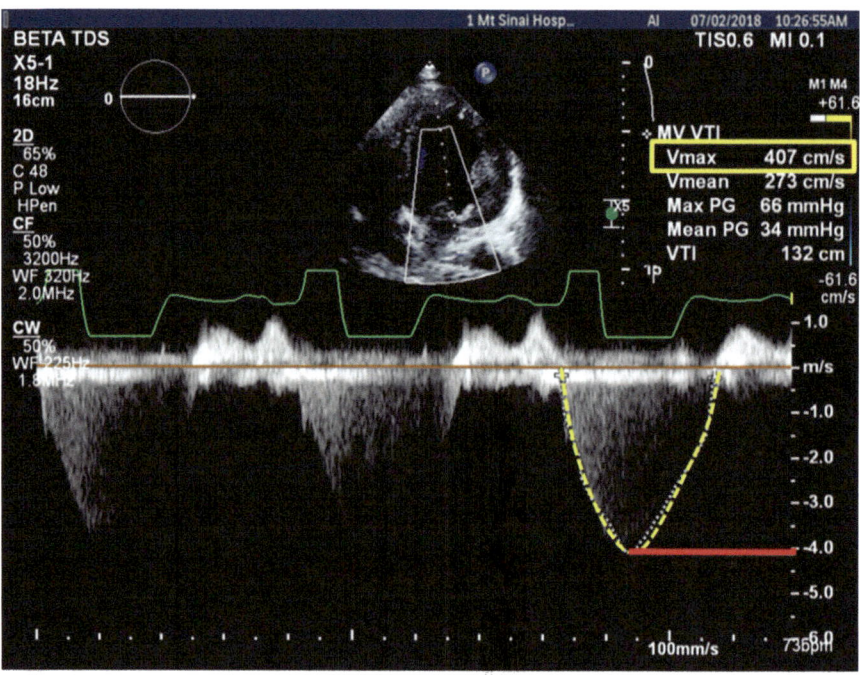

MR V_{max} = 4 m/sec

Fig. 7.1 Low-velocity mitral regurgitation (MR). Continuous-wave Doppler across the MR jet shows atypical spectral envelope; peak MR velocity is low (4 m/sec) and the shape of the jet is triangular with slow rise

7.2 Approach to an Unusual MR Tracing

7.2.1 Low Peak Velocity

- Using the simplified Bernoulli equation ($\Delta P = 4v^2$), MR peak velocity of 4 m/sec means a driving pressure gradient of $4 \times 4^2 = 64$ mmHg.
- Whenever a pressure gradient is estimated, the next two questions are:
  - Between what two chambers this pressure gradient occurs.
  - At what part of the cardiac cycle.
- Regarding MR: The calculated ΔP represents the pressure difference between the left ventricle and the left atrium during systole (ΔP_{LV-LA}).
- Reminder about typical MR gradient (see Fig. 6.3).
  - Left ventricular systolic pressure is generally ~110–120 mmHg.
  - Normal left atrial pressure is generally ~10 mmHg.
  - Pressure difference between the LV and LA during systole: $\Delta P_{LV-LA} \approx 100$ mmHg.
  - ΔP_{LV-LA} of 100 mmHg corresponds to peak velocity of 5 m/sec.
- In the presented case, the MR velocity is significantly lower—ΔP_{LV-LA} 64 mmHg.
- The calculated ΔP of 64 mmHg means that the LV systolic pressure reaches a value that is only 64 mmHg higher than LA pressure.
- Thinking about possible causes for low systolic pressure difference between the LV and LA:
  - LV systolic pressure may be low.
  - LA systolic pressure may be high.
  - Combination of the above two reasons (Fig. 7.2).
- Notably, for ΔP_{LV-LA} to be this low (64 mmHg) it is likely that both the LV systolic pressure is low, and the LA pressure is high.
- Low LV systolic pressure means low aortic pressure; high LA pressure means high filling pressures / high pulmonary venous pressures.
- A combination of low systemic blood pressure with elevated filling pressure is the hallmark of cardiogenic shock.
- In our case study—presumed numbers that can translate to ΔP of 64 mmHg are: LV systolic pressure 90 mmHg with LA pressure 25–30 mmHg.
- Thus, by looking at the peak MR velocity it appears likely that the patient is suffering from impending or manifest cardiogenic shock.

7.2.2 Abnormal MR Envelope Shape

- The second abnormal feature of this MR spectral tracing is the shape of the envelope.
- Typically, MR spectral Doppler shows a rapid rise in velocity; the initial increase in MR velocity represents the rapid rise in LV pressure at the onset of systole (starting with the isovolumetric contraction).

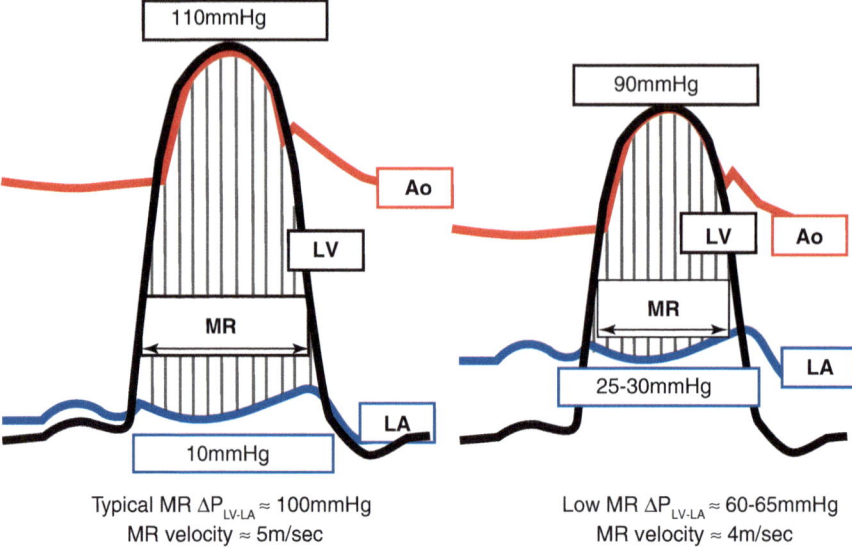

Fig. 7.2 Comparing "typical" MR to low-velocity MR. MR velocity is determined by the systolic pressure gradient (ΔP_{LV-LA}) between the left ventricle (LV) and the left atrium (LA). Low MR velocity (right diagram) implies low LV systolic pressure and high LA pressure. Low LV systolic pressure means low systolic aortic (Ao) pressure. The combination of low aortic pressure and high LA pressure is consistent with impending or manifest cardiogenic shock

- The rate of pressure increase during early systole depends on the contractile function of the left ventricle and the preload; in general, normal LV shows a rapid rise in pressure and hence rapid rise in MR velocity.
- The rate of rise in LV pressure can be quantified by calculating $\dfrac{dP}{dT}$.
- In order to calculate $\dfrac{dP}{dT}$ the MR envelope is interrogated.
  - A mark is made on the MR envelope at the point where the MR velocity is 1 m/sec.
  - At this point, $\Delta P_{LV-LA} = 4 \times 1^2 = 4$ mmHg.
  - A second mark is made on the MR envelope at the point where the velocity is 3 m/sec.
  - At this point, $\Delta P_{LV-LA} = 4 \times 3^2 = 36$ mmHg.
  - During the time between the first and second mark, the pressure in the LV has increased; the change in LV pressure is: $dP = 36 - 4 = 32$ mmHg.
  - The time difference between the two marks (dT) can be measured by the measurements software on the echo machine or on the echo viewing station.
- $\dfrac{dP}{dT}$ is calculated as $\dfrac{32}{dT(\sec)}$.

7.2 Approach to an Unusual MR Tracing

- Cutoff values for $\dfrac{dP}{dT}$ are:
  - **Normal**: Time < 27 mSec (0.027 sec), dP/dT >1200 mmHg/sec.
  - **Borderline**: Time 27–32 mSec (0.027–0.032 sec), $dP/dT \approx$ 1000–1200 mmHg/sec.
  - **Abnormal**: Time > 32 mSec (0.032 sec), dP/dT <1000 mmHg/sec.
- Looking at our case study, it can be seen (even prior to measuring) that MR jet envelop has a slow rise, less steep shape.
- For more accurate quantification (Fig. 7.3), the time it took for the MR velocity to increase from 1 m/sec to 3 m/sec is measured; in this case: 100 msec (0.1 sec).
- Thus, $\dfrac{dP}{dT} = \dfrac{32}{0.1} = 320 \, \text{mmHg}$.
- This is a very low value, consistent with low contractile function and likely increased preload.

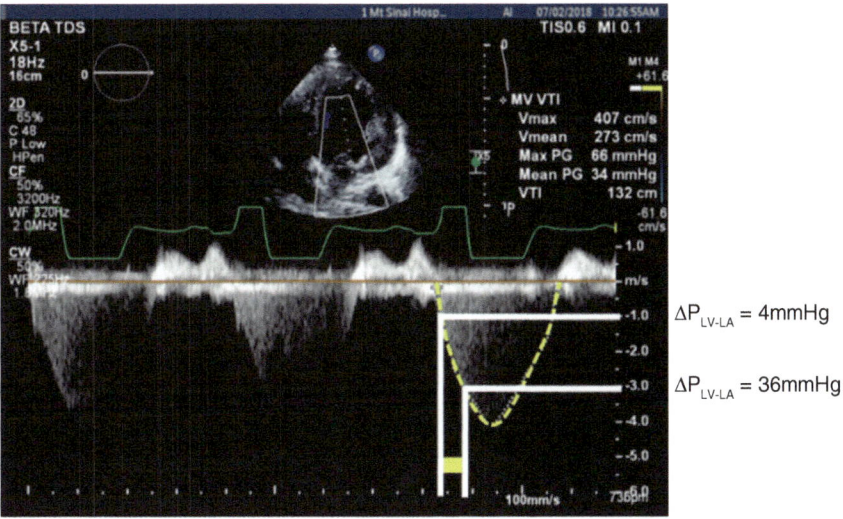

dT = 100msec = 0.1sec, dP = 36-4 = 32mmHg

Fig. 7.3 Calculating dP/dT. A mark is made on the MR envelope where the MR velocity is 1 m/sec. This marks the point of time where the pressure gradient between the left ventricle and left atrium ($\Delta P_{\text{LV-LA}}$) is 4 mmHg. A second mark is made on the MR envelope at the point where the velocity is 3 m/sec, meaning $\Delta P_{\text{LV-LA}}$ is 36 mmHg. The time difference between the two marks (dT) can be measured by the software on the echo machine or on the echo viewing station; during this time, the pressure in the LV has increased by: dP = 36–4 = 32 mmHg. dP/dT is then calculated as 32/measured dT (sec)

7.3 Summary and Final Points

- In this case study, an unusual MR envelope was detected by spectral Doppler.
- The spectral envelope was abnormal on two counts:
  - Low peak velocity.
  - Slow rise in MR velocity.
- Low peak MR velocity indicates a low systolic pressure gradient between the LV and LA, suggesting low systemic pressure with elevated filling pressure➔ impeding or manifest cardiogenic shock.
- Slow rise MR velocity indicates low pressure rise in the LV during early systole; this can be quantified by calculating $\frac{dP}{dT}$.
- In our case, the calculated $\frac{dP}{dT}$ was extremely low, suggesting that the impending shock state was due to poor LV systolic function.

Why Is This Happening Now

8

Abstract

The unusual timing of an otherwise appearing "insignificant" flow can be a clue to various underlying pathologies and should be thoroughly interrogated and investigated. While these abnormal flows may not be clinically important in and of themselves (e.g., may not cause volume overload or increase in filling pressures), understanding the cause of the abnormal timing is important, as it may clarify the underlying pathology. Diastolic mitral regurgitation is an example of such an abnormally timed event; understanding the mechanism and pressure relationships that result in diastolic MR can help shed light on the pathology causing it.

Keywords

Diastolic MR · Heart clock · Atrial relaxation

Introduction

- Echocardiography can provide important information regarding various events' timing during the cardiac cycle.
- Unusual timing of an otherwise appearing "insignificant" flow can be a clue to various underlying pathologies and should be thoroughly interrogated and explained.
- In this chapter, we will review several case studies that demonstrate how attention to timing can help shed light on cardiac function and/or dysfunction.

8.1 First-Degree Heart Block

- A 90-year-old woman was admitted for noncardiac diagnosis. Echocardiogram preformed for pre-op evaluation prior to noncardiac intervention.
- Left ventricular function was preserved with no segmental wall motion abnormalities.
- Mild-to-moderate mitral regurgitation noted.
- Color Doppler imaging, as well as spectral tracing, showed "typical" MR findings, as well as abnormally timed MR (Fig. 8.1).

8.1.1 Approach to Abnormal MR Timing

- Mitral regurgitation is generally a systolic phenomenon.
- As the left ventricular contraction starts, the pressure in the LV rises above the pressure in the LA and the mitral valve closes (first heart sound).
- If the mitral valve is incompetent, MR starts as soon as the pressure in the LV exceeds the pressure in the LA and lasts throughout mechanical systole; as discussed earlier, MR is typically a "holosystolic" event meaning it lasts throughout isovolumetric contraction, ejection time, and isovolumetric relaxation.
- In the case shown, a diastolic color Doppler frame is seen and a blue flow (away from the transducer) is seen during diastole. Note that even though the flow is mitral regurgitation, from LV to LA, it is mainly coded in laminar blue color, suggesting low-velocity flow.
- The corresponding spectral Doppler image shows the typical systolic MR envelope, however, in addition, a low velocity, late diastolic MR is seen.
    - Note that it is important to verify that the late diastolic flow seen on the spectral Doppler is in fact diastolic MR; continuous-wave Doppler does not allow

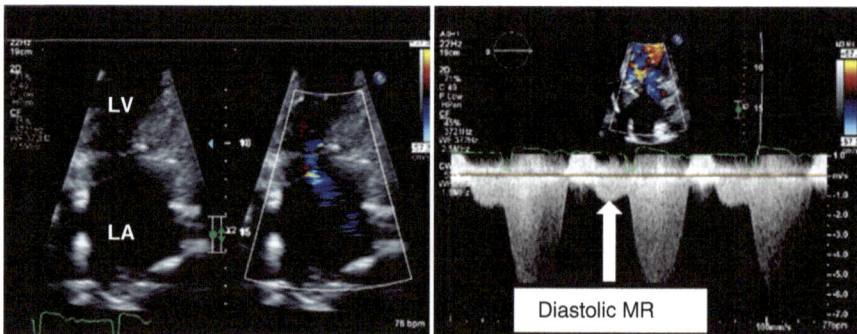

Fig. 8.1 Diastolic mitral regurgitation (MR) in first-degree heart block. Late diastolic frame from color apical four-chamber view (*left panel*) showing laminar flow from the left ventricle (LV) into the left atrium (LA). Corresponding spectral Doppler (*right panel*) showing the typical, high velocity, systolic MR as well as late diastolic, low velocity "diastolic MR"

8.1 First-Degree Heart Block

  for localization of the measured flow and hence may record flow from other areas along the path of the beam.
  - Combing information obtained by spectral tracing with 2D and color Doppler imaging is essential for proper interpretation.
- The added flow that is seen immediately preceding the typical MR envelops can be characterized as:
  - Late diastolic.
  - Seen with every beat.
  - Low velocity (peak velocity 1.5 m/sec).
- In addition, looking closely at the spectral tracing, a forward flow is seen immediately prior to the low-velocity MR signal.
- Looking at the ECG tracing on the spectral tracing, the QRS is seen at the beginning of the typical MR jet, low amplitude T wave is seen in early diastole, followed shortly thereafter by a P wave, which in turn is followed by a very long PR interval.
- The patient's 12-lead ECG is shown (Fig. 8.2), confirming the presence of first-degree atrioventricular block (with PR interval measured at 432 mSec).
- Diastolic MR commonly accompanies first-degree AV block (Fig. 8.3):
  - Atrial contraction is followed by atrial relaxation.
  - Atrial relaxation causes the left atrial pressure to drop.
  - In the presence of normal atrioventricular (AV) conduction, ventricular pressure shows a small "bump" at end diastole (due to the increased volume from the atrial contraction), which is then immediately followed by the onset of

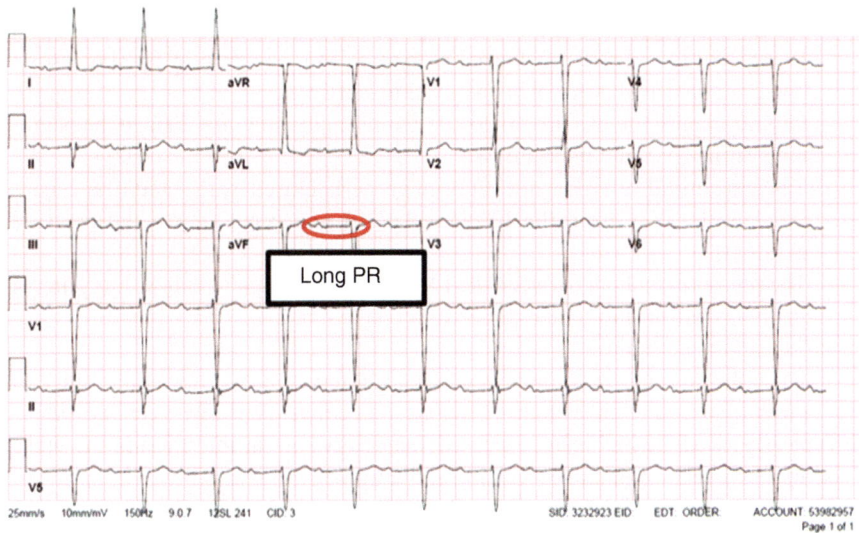

Fig. 8.2 ECG. Twelve lead ECG from the patient presented showing normal sinus rhythm with first-degree AV block (PR interval measure 432 mSec)

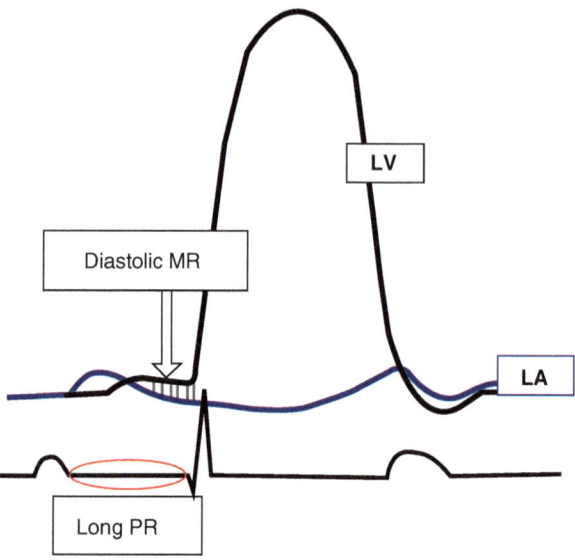

Fig. 8.3 Pressure tracings in first-degree heart block. Left ventricular (LV) and left atrial (LA) pressures are shown. Atrial contraction is followed by atrial relaxation, causing the left atrial pressure to drop. In the presence of first degree AV block, ventricular systole is delayed to start such that LA pressure can drop below LV pressure. Since LV pressure exceeds LA pressure, MR can occur; notably, the pressure gradient between the LV and LA at this point is low (since rapid rise in LV pressure is late to start), hence the MR is of low velocity

  ventricular systole with rapid increase in ventricular pressure and closure of the mitral valve.
 – When first-degree AV block is present, atrial contraction causes a small increase in LV pressure and atrial relaxation causes a drop in atrial pressure, however, ventricular systole is delayed to start.
 – Left atrial pressure can drop below left ventricular pressure, **before** the beginning of the LV rapid pressure rise.
 – Since the mitral valve has not yet been properly closed by the start of mechanical systole, flow from the LV into the LA can occur (down the pressure gradient).
 – The pressure gradient between the LV and LA at this point is **low**; mechanical systole is late to start (due to the first-degree AV block) and LV pressure is still "diastolic" pressure.
 – Since the pressure gradient between the LV and the LA is low at this stage of the cardiac cycle, the MR that happens at late diastole is **low-velocity** MR.
• Typically, diastolic MR is not a hemodynamically significant lesion in the sense of volume overload; the diastolic mitral regurgitant volume is low.
• The significance of late diastolic MR is that it signifies a discrepancy between the timing of atrial pressure dropping below ventricular pressure and onset of mechanical systole.
• While first-degree heart block is a common cause for such discrepancy, it is not the only cause (see Chap. 10).

8.2 High Degree Heart Block

- A 78-year-old man, admitted after a syncope event.
- Describes sudden onset of light-headedness followed by immediate collapse; this followed a two weeks period of weakness and general fatigue.
- Past medical history significant for hypertension only.
- When seen in the emergency department, the patient was already awake, alert, and oriented, afebrile, HR ~ 30/min, BP 116/74.
- Initial workup included 12 lead ECG (Fig. 8.4) and an echocardiogram (Fig. 8.5).

8.2.1 Approach to Abnormal MR Timing

- Similar to the prior case study, a color Doppler diastolic frame is seen, showing diastolic laminar flow from LV into LA (diastolic mitral regurgitation).
- The corresponding spectral tracing shows a single envelope of typical systolic MR, with several diastolic envelopes depicting several diastolic MR occurrences.
- Note that prior to every diastolic MR envelope, a small forward flow (LA into LV) is seen.
- Looking at the ECG tracing on the images as well as the 12-lead ECG that was obtained, a complete heart block is identified.
  - There is a clear atrioventricular dissociation with atrial rhythm of approximately 100/min, with a ventricular escape rhythm of approximately 30/min.
- Every atrial contraction results in forward flow from LA to LV (as seen on the spectral tracing). The increased LV volume causes a slight increase in LV pressure.
- In addition, following atrial contraction, atrial relaxation ensues, causing a drop in the left atrial pressure.

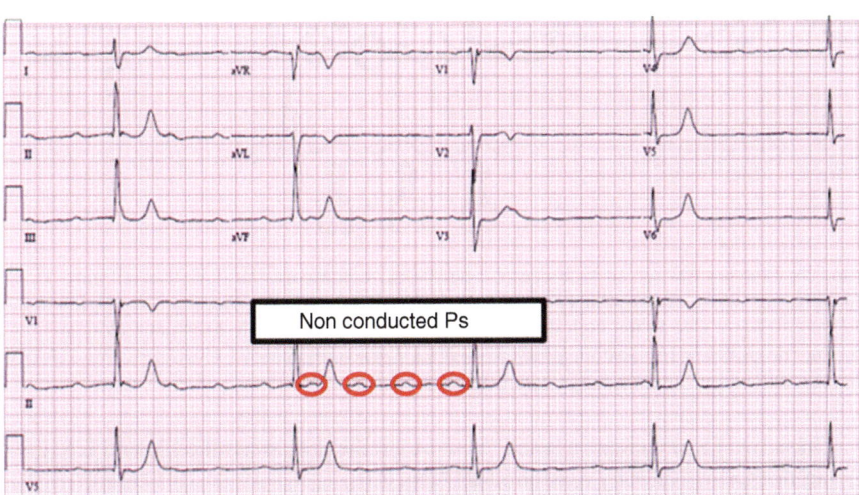

Fig. 8.4 ECG. Twelve lead ECG from the second patient presented showing high degree AV block. There are multiple non-conducted P waves and there is clear AV dissociation

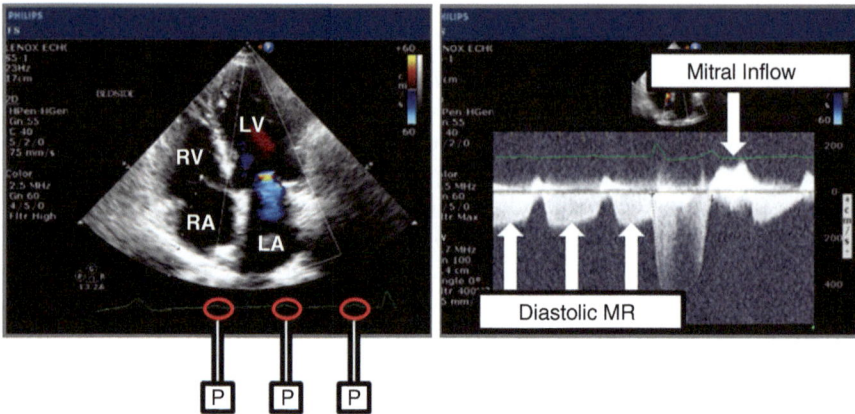

Fig. 8.5 Diastolic mitral regurgitation (MR) in high degree AV block. Color Doppler diastolic frame from apical four-chamber color view (left panel) showing diastolic laminar flow from left ventricle (LV) into left atrium (LA). The corresponding spectral tracing (right panel) shows a single envelope of typical systolic MR, with several diastolic envelopes depicting several diastolic MR occurrences. Notably, prior to every diastolic MR envelope, a small forward flow (LA into LV) is seen (RA—right atrium, RV—right ventricle)

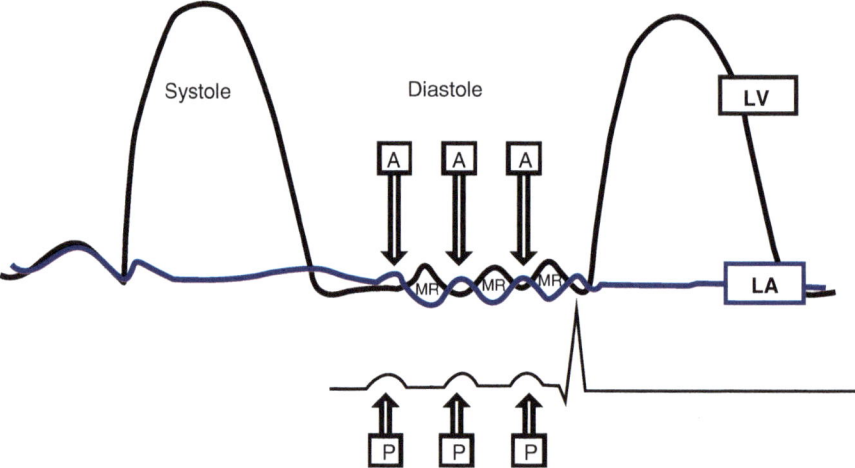

Fig. 8.6 Pressure tracings in high degree heart block. Left ventricular (LV) and left atrial (LA) pressures are shown. Every atrial contraction is followed by atrial relaxation, causing the left atrial pressure to drop. Due to the high degree AV block, ventricular systole does not follow and LA pressure drops below LV diastolic pressure. Since LV pressure exceeds LA pressure, low-velocity diastolic MR is seen after each non-conducted atrial activation

- These two factors create a small pressure gradient between the LV and LA, giving rise to low-velocity diastolic MR (Fig. 8.6).
- When ventricular systole does occur, the LV systolic pressure rises rapidly; a "typical" high-velocity MR is then seen.

- Following ventricular systole, diastole starts with LV relaxation; when LV pressure drops below LA pressure, the mitral valve opens and rapid ventricular filling starts.
- As diastole continues, atrial contractions (which are not followed by ventricular systole) take place, again giving rise to several diastolic MR occurrences.

8.3 The Missing Atrial Contraction

- A 60-year-old man, presented to the emergency room with chest pain.
- ECG and echocardiogram were obtained.

8.3.1 Approach to Abnormal MR Timing

- Similar to the prior case studies, a color Doppler diastolic frame is seen, showing diastolic laminar flow from LV into LA (diastolic mitral regurgitation) (Fig. 8.7).
- The spectral Doppler tracing shows a very unusual pattern.
- Following the T wave (during early diastole) diastolic inflow from LA into LV is seen (during the rapid filling phase).
- At late diastole, there is a smaller forward flow corresponding to the atrial contraction.
- Note that in between these two envelopes there is a low-velocity backward flow (LV to LA, i.e., early diastolic mitral regurgitation).
- Although neither the ECG tracing on the image nor the spectral Doppler shows an atrial contraction proceeding this diastolic MR flow, the mere presence of this flow is the clue that there was a "**hidden**" atrial contraction.

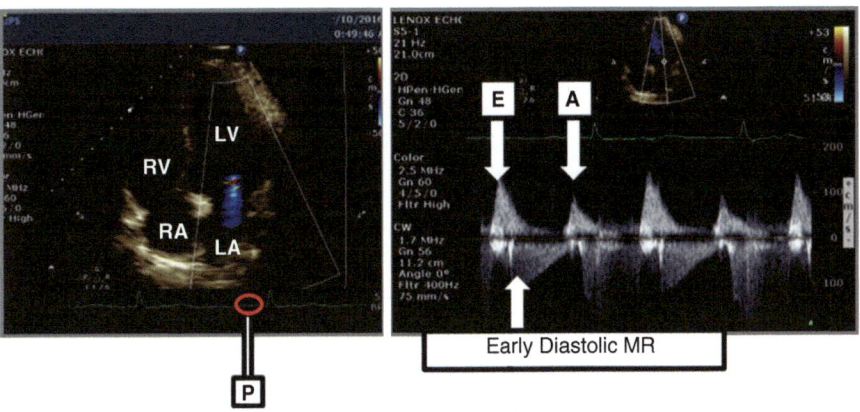

Fig. 8.7 Early diastolic mitral regurgitation (MR). Color Doppler diastolic frame from apical four-chamber color view (left panel) showing diastolic laminar flow from left ventricle (LV) into left atrium (LA). The corresponding spectral tracing (right panel) shows the typical mitral inflow (E and A waves) along with an early diastolic envelope of low-velocity MR (RA—right atrium, RV—right ventricle)

- The "hidden" atrial contraction occurred early in diastole, simultaneously with the rapid filling phase, and hence could not be picked up by the spectral Doppler.
- The hidden atrial contraction was followed by atrial relaxation which made the LA pressure drop below the LV pressure, thus allowing for diastolic mitral regurgitation to occur (Fig. 8.8).
- ECG showed second-degree heart block, with non-conducted P waves happening in early diastole (Fig. 8.9).

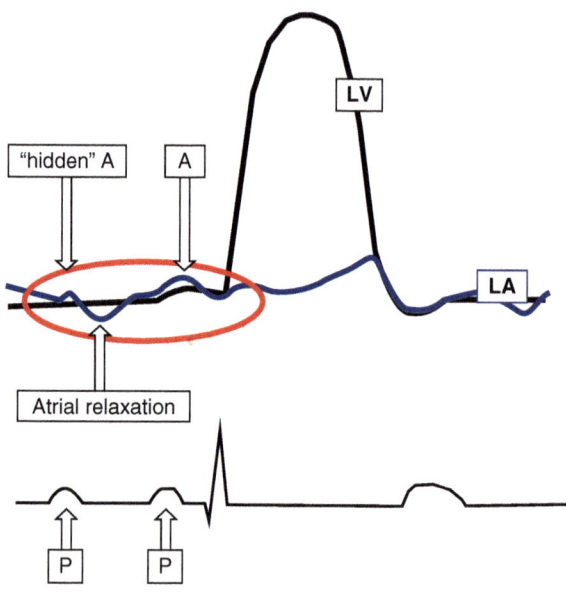

Fig. 8.8 Pressure tracings with "hidden" atrial contraction. Inflow from left atrium (LA) into left ventricle (LV) starts at early diastole during the rapid filling phase. Simultaneously, a "hidden" atrial contraction occurs, further contributing to the forward LA-LV flow. The hidden atrial contraction is followed by atrial relaxation which causes the LA pressure to drop below the LV pressure, thus allowing for early diastolic, low-velocity mitral regurgitation

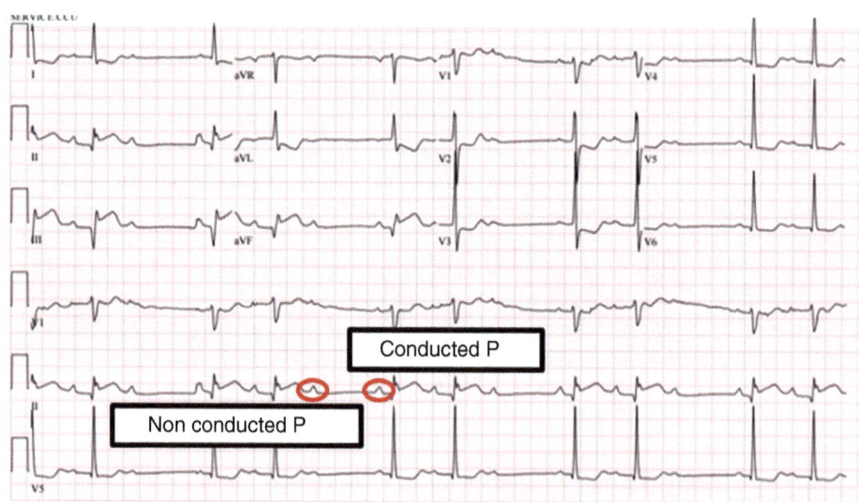

Fig. 8.9 ECG. 12 lead ECG from the third patient presented showing second-degree heart block, with non-conducted P waves happening in early diastole. Additionally, inferior Q waves and ST elevations can be seen, suggesting the conduction system abnormality was due to acute coronary syndrome

- In addition, inferior Q waves and ST elevations can be seen, suggesting that the cause for the conduction system abnormality was an acute coronary syndrome.

8.4 Summary and Final Points

- The unusual timing of intracardiac flows can occasionally be identified on Doppler images and spectral tracings.
- While these abnormal flows may not be clinically important in and of themselves (e.g., may not cause volume overload or increase in filling pressures), understanding the cause of the abnormal timing is important, as it may shed light on the underlying pathology.
- In this chapter, we examined three cases where diastolic MR was seen due to atrioventricular conduction diseases (first-, second-, and third-degree heart blocks).
- In all these cases, the combination of atrial contraction causing a slight increase in left ventricular pressure as well as atrial relaxation causing a slight decrease in left atrial pressure resulted in small diastolic pressure gradient between the LV and LA, allowing for low-velocity, diastolic MR.
- It is important to remember that there may be other pathologies that result in diastolic pressure gradient between LV and LA and it is essential to identify them correctly (see Chap. 10).

Extreme Pulsus Alternans

Abstract

Timing of events, as well as the pattern of occurrences, are important details to pay attention to. Occasionally, a discrepancy can be found between electrical and mechanical activity of the heart, and echocardiography can be a tool to identify such inconsistency. Moreover, there can be a discrepancy between the mechanical activation of the heart, and the efficiency of the heart (i.e., the production of cardiac output), and echocardiography can play a central role in diagnosing such a situation. Combining information from ECG tracing, 2D and tissue Doppler assessment of contractility, and spectral tracing across the LVOT can provide evaluation of the electrical activation of the heart, mechanical activation of the heart, and the ultimate "success" of these activation efforts in producing cardiac output.

Keywords

Systolic dysfunction · Cardiomyopathy · Pulses alternans · Cardiac output · Cardiac index

Introduction

- Timing of events, as well as pattern of occurrences are important details to pay attention to.
- In this chapter, we will review a case study that demonstrates how attention to event pattern can help clarify pathophysiology of a clinical syndrome.

9.1 Case Presentation

- A 68-year-old man with a known cardiomyopathy with systolic dysfunction, presented to the emergency department with weakness, fatigue, and shortness of breath.
- His pulse was weak to palpation, blood pressure difficult to measure but appeared to be low, approximately 80/45, tachypnea to ~25/min noted, and his extremities were cold and clammy.
- His echocardiogram showed dilated left and right ventricles with biventricular systolic dysfunction; LV ejection fraction was 15%.
- Tissue Doppler tracing from the lateral mitral annulus and pulse Doppler of the LVOT were acquired (Fig. 9.1).
- Pulse Doppler at the LVOT is shown (Fig. 9.2).
  - Ventricular trigeminy is seen on the ECG tracing on the images; every third beat is a ventricular premature contraction.
  - The tissue Doppler tracing shows an S' wave associated with each electrical activity, confirming mechanical activation of the LV with every electrical activation.
  - However, the LVOT pulse Doppler shows a different pattern; a signal is obtained only once every three beats.
  - Only the sinus beat that follows the PVC produces an LVOT envelope; the following sinus beat, as well as the PVC do not produce any LVOT tracing.
  - The presence of LVOT tracing represents flow across the LVOT, meaning a stroke volume is created; lack of LVOT tracing means no flow across the LVOT, which means no effective stroke volume.
  - There is a discrepancy between the rate of LV electrical and mechanical activation (approximately 75/min) and the rate of stroke volume production (approximately a third of that, ~25/min).

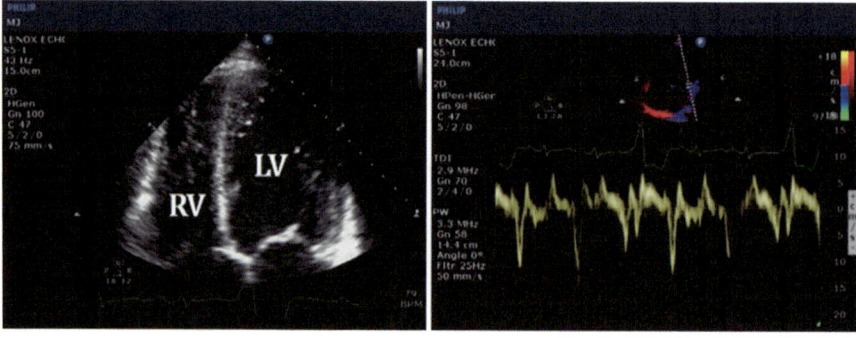

Fig. 9.1 2D image and tissue Doppler tracing. Apical four-chamber view (*left panel*) concentrating on the left ventricle (LV) and right ventricle (RV). The LV appears dilated. Notably, the RV is apex forming and has a lateral diameter nearly equal to that of the LV—both suggestive of RV dilatation. Tissue Doppler tracing from the lateral mitral annulus (*right panel*) showing the classic systolic S' wave and diastolic waves

9.1 Case Presentation

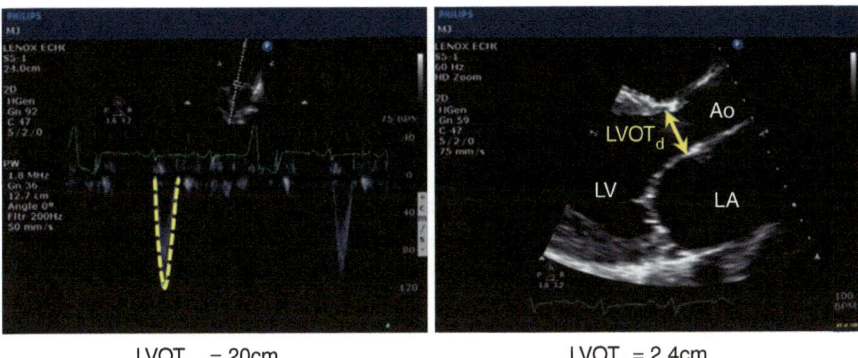

LVOT$_{VTI}$ = 20cm   LVOT$_d$ = 2.4cm

Fig. 9.2 LVOT assessment. Pulse wave Doppler tracing from the LVOT (*left panel*) showing spectral envelope (LVOT$_{VTI}$) only with every third QRS. No discernable flow can be seen with the other QRS complexes. Right panel showing 2D parasternal long axis image, for LVOT diameter (LVOT$_d$) measurement. Utilizing the LVOT$_d$ and LVOT$_{VTI}$, stroke volume can be calculated (for those beats that create a stroke volume)

- Using VTI and cross-sectional area, the stroke volume can be calculated.
- In this case: LVOT$_d$ = 2.4 cm, LVOT$_{VTI}$ = 20 cm.
- SV = LVOT$_{CSA}$ ×LVOT$_{VTI}$=$\pi \times 1.2^2 \times 20$ = 90cc
- While this may appear as adequate stroke volume, calculating the cardiac output reveals a different assessment:
- CO = SV × HR
- Important to remember that the **effective** heart rate, in this case, is **different** than the ECG or mechanical heart rate.
- Thus, for calculating the correct cardiac output (or index) the effective heart rate must be used; meaning CO = SV × HR$_{effe}$=90 × 25 = 2.25 lit/min.
- The patient's BSA was 1.97 m^2 → cardiac index 1.14 lit/min/m^2.
- Normal cardiac output is approximately 5–6 lit/min, cardiac index 2.5–4 lit/min/m^2.
• Despite the "normal" calculated stroke volume, the effective cardiac output/cardiac index was very low.
• Thinking about the underlying mechanism:
  - Pulses alternans can occasionally accompany left ventricular systolic dysfunction.
  - Generally, it manifests as alternating "strong" and "weak" beats, even in the presence of a normal heart rate.
  - The precise mechanism remains not entirely clear; suggested explanations include:
    Changes in preload (and afterload) causing LV contractile changes as predicted by the Frank-Starling relation.
    Changes in LV contractility due to changes in the sarcoplasmic calcium pump.

- Our case study shows an "extreme" example of pulses alternans; the pulse alternates between presence of a beat, and a complete absence of a beat.
- The presence of the ventricular trigemini further amplifies the phenomenon:
  - The PVC timing and efficiency are such that it cannot produce mechanically adequate contraction to create a stroke volume.
  - The result is that the preload for the following beat is increased, which in turn increases the contractility and the stroke volume.
  - Since the post-PVC beat emptied the heart efficiently, the next beat's preload is inadequate for stroke volume generation; thus, no effective LVOT flow is generated.
  - Given the trigemini pattern, the following beat is a PVC, which is poorly timed and mechanically inadequate for SV production → this creates a large preload for the next beat → SV generated on the next sinus beat.

9.2 Summary and Final Points

- In this case study, the LVOT pulse-Doppler pattern was unique as flow could only be seen every third beat.
- This created an "extreme" pulses alternans situation, where effective stroke volume was generated only every third electrical beat.
- It is important to realize that even though the heart **did** contract with every electrical activation (as evidenced by the tissue Doppler tracing showing S' wave following each QRS), two out of every three LV contractions were essentially ineffective and produced no forward stroke volume.
- When calculating the patient's cardiac output/index, the effective heart rate must be considered.
- From all the above, it appears that the patient's presenting symptoms were due to decompensated heart failure with extremely low cardiac output, as least partially due to extreme pulses alternans, worsened by the presence of ventricular trigemini.

Is This AI an Emergency

10

Abstract

Aortic insufficiency (AI) is frequently encountered in echocardiography; regardless of its cause (e.g., valvular, aortic root pathology, etc.), accurate quantification of AI severity is essential for prognostication and treatment planning. Special attention should be paid to parameters that suggest hemodynamically significant AI with increased LV end-diastolic pressure; these may be signs of impending hemodynamic collapse and may necessitate prompt intervention. While often severity and hemodynamic significance go hand in hand, there can be physiologic reasons or technical reasons for the AI to appear less than severe, yet with signs of significant hemodynamic consequences. The importance of identifying these signs cannot be overemphasized; timely intervention may be lifesaving and echocardiographic findings may precede clinical deterioration.

Keywords

Aortic regurgitation · Aortic insufficiency · Regurgitant volume · PISA method Vena contracta · Premature closure of the mitral valve · Diastolic MR · Diastolic reversal of flow · Pressure half time · AI end-diastolic velocity

Introduction

- Aortic insufficiency (AI) is frequently encountered in echocardiography.
- Accurate quantification of AI severity is of paramount importance for proper prognostication and treatment planning.
- In addition to quantification, it is important to pay attention to signs that can point to the hemodynamic significance of the AI.
- While often severity and hemodynamic significance go hand in hand, the importance of identifying the hemodynamically significant signs cannot be

overstated; timely intervention may be lifesaving and echocardiographic findings may precede clinical deterioration.
- In this chapter, we will review a few techniques by which to quantify AI severity and we will emphasize parameters that indicate hemodynamically significant AI.

10.1 AI Quantification

- Aortic insufficiency happens when blood spills back through the aortic valve into the LV, during diastole.
- AI can be caused by valvular pathologies (e.g., leaflet abnormalities like endocarditis or bicuspid valve) or aortic root pathologies (e.g. aortic root aneurysm or aortic dissection).
- Similar to mitral regurgitation, the severity of AI is defined by the volume (or fraction) of blood that regurgitates back into the left ventricle during diastole.
- Other parameters used for quantification serve as surrogates for the severity; pitfalls and shortcomings should be acknowledged in order to accurately quantify AI severity.
- The AI volume (regurgitant volume [RV]) can be estimated quantitatively and qualitatively by several techniques:

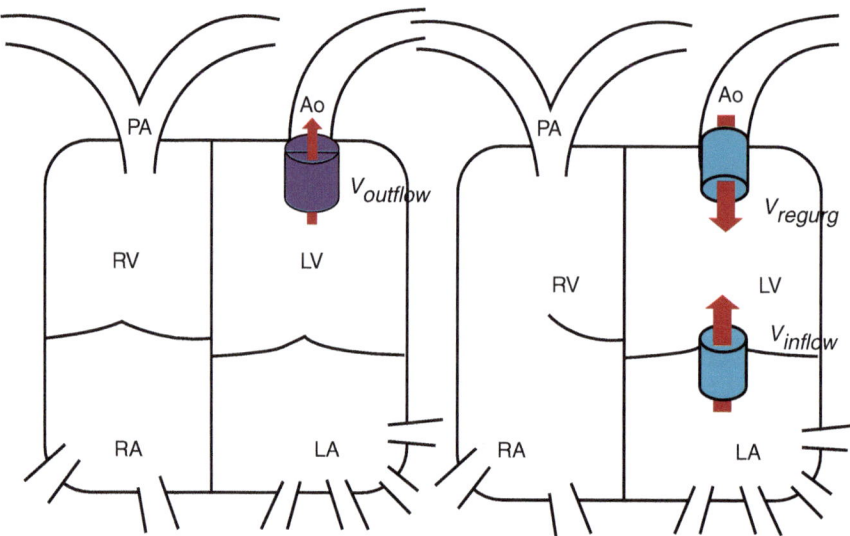

Fig. 10.1 Volumetric calculation of aortic regurgitant volume (RV). The mitral inflow volume (V_{inflow}) can be calculated by measuring the mitral annulus diameter and VTI. The outflow volume from the left ventricle ($V_{outflow}$), which includes the regurgitant volume can be calculated by measuring the LVOT diameter and VTI. The difference between these two volumes ($V_{outflow}$-V_{inflow}) is the RV (Ao—aorta, LA—left atrium, LV—left ventricle, PA—pulmonary artery, RA—right atrium, RV—right ventricle)

10.1 AI Quantification

- Volumetric calculation based on the conservation of mass principle (Fig. 10.1):

  In isolated AI (e.g., no shunts and MR), the flow volume across the LVOT can be compared to the inflow across the mitral annulus.

  The flow volume across the LVOT is larger than the inflow into the LV, as it includes the AI regurgitant volume.

  RV = LVOT outflow − MV annular inflow.

  $RV = \pi r_{LVOT}^2 \times LVOT\ VTI - \pi r_{annulus}^2 \times MV_{annulus}\ VTI$.

  $$\text{Regurgitant Fraction} = \frac{RV}{LVOT\ outflow}.$$

- PISA method to calculate the aortic regurgitation effective regurgitant area and the RV (using same principles as described for MR; see Chap. 5).
- The width of the AI jet relative to the LVOT can be assessed (Fig. 10.2); this is best estimated on the parasternal long-axis view; severe AI takes up more than 65% of the LVOT.

  Important to note that eccentric AI jets (e.g., prosthetic paravalvular leaks and bicuspid aortic valves) may appear misleading as the jet can track along the LVOT wall.

  Notably, jet length (i.e., how far back into the LV the jet extends) does not seem to be a reliable quantifying parameter.

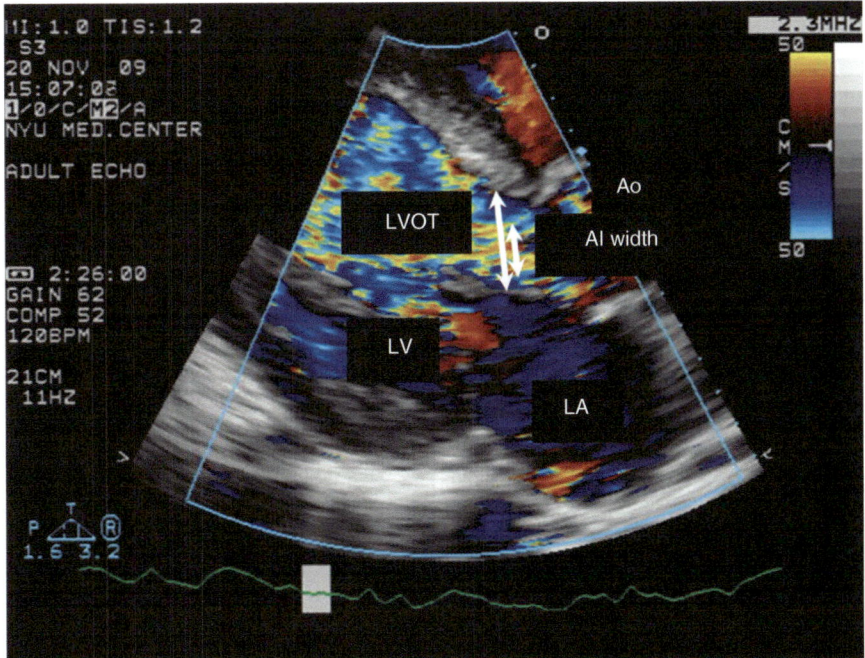

Fig. 10.2 Aortic insufficiency (AI) jet width. Diastolic frame from parasternal long-axis view showing a very wide AI jet. AI jet width relative to the LVOT can be used to quantify AI severity; severe AI takes up more than 65% of the LVOT (Ao—aorta, LA—left atrium, LV—left ventricle)

- Other findings that can suggest high regurgitant volume include dense AI spectral Doppler envelope, as well as dense forward flow spectral Doppler envelope with high peak/mean gradient (despite the absence of aortic stenosis).
- In addition to calculating or estimating regurgitant volume, the AI effective regurgitant orifice area (EROA) can be evaluated.
  - PISA method can be used to calculate the AI EROA.
  - Vena contracta can be measured as a surrogate for the EROA.
- Cutoffs for AI gradation are shown in Table 10.1.
- Note that while cutoffs for severe AI regurgitant volume and regurgitant fraction are the same as for MR (60 ml / 50% respectively), the cutoffs for EROA and VC are lower than those for MR.
- The difference relates to the event timing:
  - AI is a diastolic phenomenon, MR happens during systole.
  - Diastole is typically longer; approximately twice as long as systole.
  - As discussed above, severity is ultimately determined by the regurgitant volume (or fraction); if more than 60 ml of blood (or more than 50% of the cardiac output) spills backward (either to LV or LA), the regurgitant lesion is severe.
  - Since diastole is longer, a significant volume of blood can regurgitate back into the LV, even through a smaller aortic EROA; on the flip side, in order for

Table 10.1 Parameters for aortic insufficiency (AI) gradation

	Mild	Moderate	Severe
Vena contracta (cm)	< 0.3	0.3–0.6	> 0.6
Jet width/LVOT width (%)	< 25	25–64	≥ 65
EROA (cm^2)	< 0.1	0.1–0.3	≥ 0.3
Regurgitant volume (ml)	< 30	30–59	≥ 60
Regurgitant fraction (%)	< 30	30–49	≥ 50

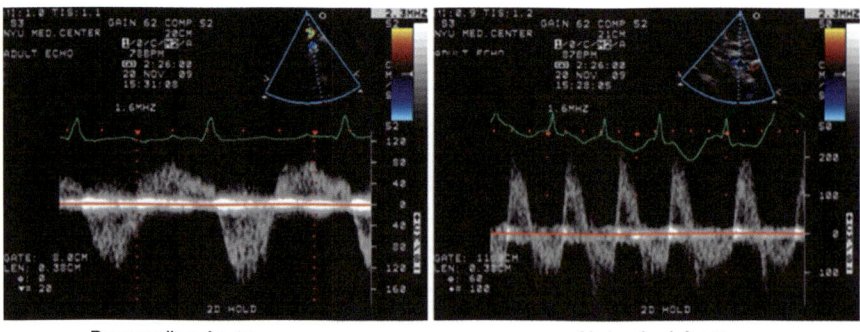

Descending Aorta | Abdominal Aorta

Fig. 10.3 Diastolic reversal of flow in the aorta. Holodiastolic flow reversal in the aorta is a sign of significant AI. Prolonged (holodiastolic as opposed to early) reversal of flow is more specific for severe AI. Additionally, reversal of flow in the abdominal aorta (*right panel*) is more specific than in the descending thoracic aorta (*left panel*)

- the same volume of blood to spill back to the LA during systole (shorter time span), the mitral EROA has to be larger.
- Diastolic reversal of flow in the descending thoracic aorta or even abdominal aorta can be seen with severe AI (Fig. 10.3); these are the echocardiographic correlates to the peripheral signs that can be seen on physical examination of patients with significant AI.
- The effect of aortic insufficiency on the heart (LV/LA dilatation) as well as the morphology of the aortic valve and aortic root should also be taken into account when assessing the AI severity.
- Yet, it is important to remember that factors other than severity only determine the hemodynamic significance of AI; the cardiac response can vary dramatically based on the acuity of the aortic insufficiency and the LV compliance.

10.2 AI Hemodynamics

- The clinical response to significant AI varies; valve disease that develops gradually can be well tolerated for a long time, whereas acute illness (e.g., infective endocarditis, aortic dissection) can cause rapid collapse and respiratory failure.
- Left ventricular compliance is an important determinant in the response to AI.
- LV compliance (C) is defined as the change in LV volume (ΔV) divided by the change in pressure (ΔP); $C = \dfrac{\Delta V}{\Delta P}$.
- Often the reciprocal of the LV compliance is referred to; the change in pressure for any given change in volume; this describes the LV "stiffness."
- Compliance (or stiffness) is determined by structural characteristics of the heart (e.g., muscle fiber orientation, presence of connective tissue) as well as contractile/relaxation state of the heart.
- AI that develops over time, allows the heart to respond by dilating and increasing LV compliance; thus, at least initially, the LV can accommodate the increased volume without a significant increase in the LV diastolic pressure.
- However, acute AI can occur on an "unprepared" LV with normal or reduced compliance which can result in severe elevation of the LV diastolic pressure.
- Certain situation can be especially grave when acute AI develops.
    - When significant paravalvular AI develops in the setting of transcutaneous aortic valve replacement (TAVR), the LV is particularly unprepared to accommodate the AI.
    - TAVR is typically performed for the treatment of aortic stenosis (AS); the cardiac response to long-standing AS—pressure overload on the heart—is LV hypertrophy, which results in decreased LV compliance.
    - If significant PVL develops during TAVR procedure, it happens acutely, on a ventricle that is not prepared for an increased volume; in fact it is prepared as poorly as possible for an increased volume, due to the preceding long-standing pressure overload.

- In this situation, acute significant AI can cause rapid deterioration with respiratory collapse.
- Certain echocardiographic parameters can point to elevated diastolic pressure and should serve as warning signs for unfavorable course.
  - AI spectral Doppler pattern—Pressure half time.

    The pressure gradient driving aortic insufficiency is the diastolic pressure difference between the aorta and the left ventricle ($\Delta P_{Ao\text{-}LV}$).

    As long as there is a pressure difference between the aorta and the LV, regurgitation from the aorta to the LV persists.

    A compliant ventricle (as seen in chronic severe AI) can accommodate the increased diastolic volume with only a small increase in LV diastolic pressure (Fig. 10.4); however, if the LV is non-compliant, the increased LV diastolic volume will cause a rapid rise in the LV diastolic pressure (Fig. 10.5).

    The rapid rise in LV pressure, along with the rapid decrease in aortic pressure, causes a quick decline in the pressure gradient between the aorta and the left ventricle, which manifests on the spectral Doppler as a steep deceleration of the AI envelope.

    Quantitatively, the slope can be measured or the pressure half time (PHT) can be measured (the time it takes for $\Delta P_{Ao\text{-}LV}$ to drop to half of its initial value); PHT < 200 mSec suggests a rapid increase in LV diastolic pressure. The rapid deceleration of the AI jet, along with the low-end diastolic velocity (see below) are the echocardiographic correlates of the short AI murmur that often accompanies acute AI; it is not uncommon to miss the diastolic murmur of acute severe AI on physical examination, precisely

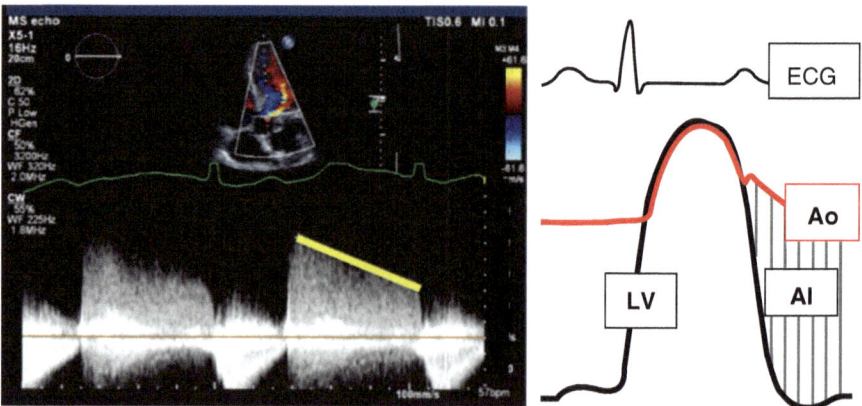

Fig. 10.4 Chronic aortic insufficiency (AI) spectral Doppler and pressure tracing. A compliant ventricle (as seen in chronic severe AI) can accommodate the increased diastolic volume with only a small increase in LV diastolic pressure. Thus, the gradient between the aorta (Ao) and the left ventricle (LV) persists throughout diastole and the deceleration slope of the AI spectral Doppler (*yellow line*, left panel) is relatively flat

10.2 AI Hemodynamics

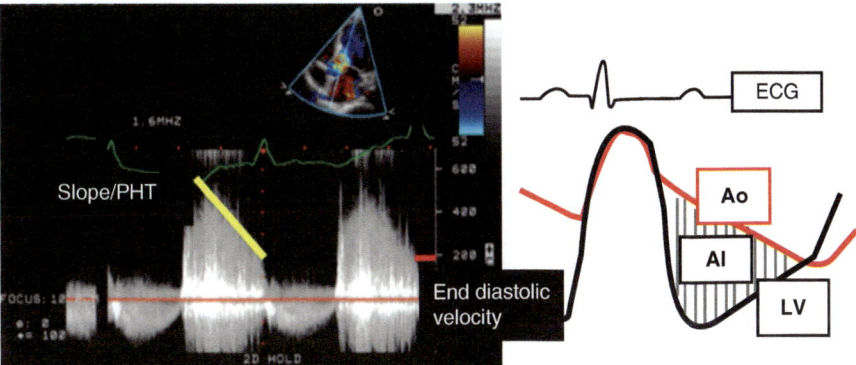

Fig. 10.5 Acute aortic insufficiency (AI) spectral Doppler and pressure tracing. Significant AI on a noncomplaint left ventricle (LV) results in a rapid rise in LV diastolic pressure. The aortic (Ao) and left ventricular (LV) pressures equilibrate fast, giving rise to a steep deceleration slope (*yellow line*, left panel) of the AI spectral tracing. In addition, since the Ao and LV pressures nearly equilibrate by end diastole, the AI end-diastolic velocity is low (*red line*, left panel)

because it can be a relatively short one; important to remember that short AI murmur may not imply mild / insignificant AI.
- AI spectral Doppler pattern—AI end-diastolic velocity.
  The end-diastolic velocity of the AI spectral tracing can also provide clinically important information.
  The AI end-diastolic velocity is determined by the pressure gradient between the aorta and the left ventricle at end diastole.
  Under normal circumstances—If the diastolic aortic pressure is relatively preserved (~70 mmHg) and the left ventricular end-diastolic pressure (LVEDP) is normal (~10 mmHg), end-diastolic ΔP_{Ao-LV} is approximately 60 mmHg, corresponding to end-diastolic velocity of 3.8 m/sec.
  With a noncompliant ventricle, LV diastolic pressure rises quickly during diastole; in severe cases, the LV pressure can increase such that it nearly equals the end-diastolic aortic pressure.
  In the case shown, the AI end-diastolic velocity is only 2 m/sec, corresponding to a ΔP_{Ao-LV} of 16 mmHg; such low-pressure gradient is the result of significantly elevated LVEDP, as well as very low aortic end-diastolic pressure, both signs of impending (or clinically manifest) hemodynamic collapse.
- Mitral valve m-mode—premature closure of the mitral valve.
  Under normal circumstances, the mitral valve closes immediately following the QRS (Fig. 10.6), after mechanical systole has started and the rapid rise in LV pressure has commenced.
  As explained above, hemodynamically significant AI causes a steep increase in LV diastolic pressure.
  If the LV diastolic pressure rises enough, it can exceed the left atrial pressure, causing premature closure of the mitral valve (Fig. 10.7).

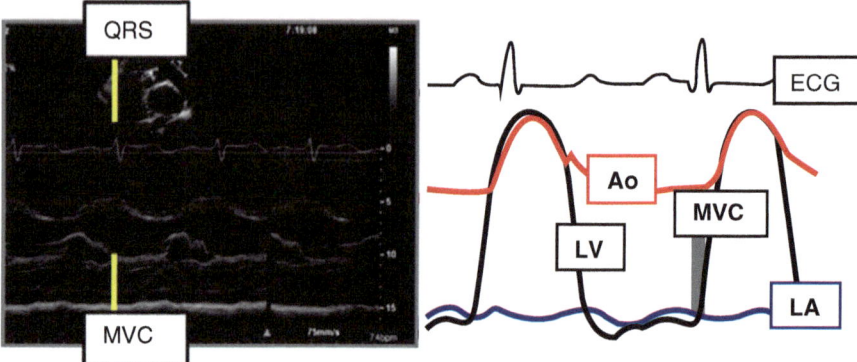

Fig. 10.6 Normal timing of mitral valve closure (MVC). Normally, the mitral valve closes when left ventricular (LV) pressure rises above left atrial (LA) pressure, at the onset of systole. This happens simultaneously or immediately following the QRS (*yellow lines*, left panel)

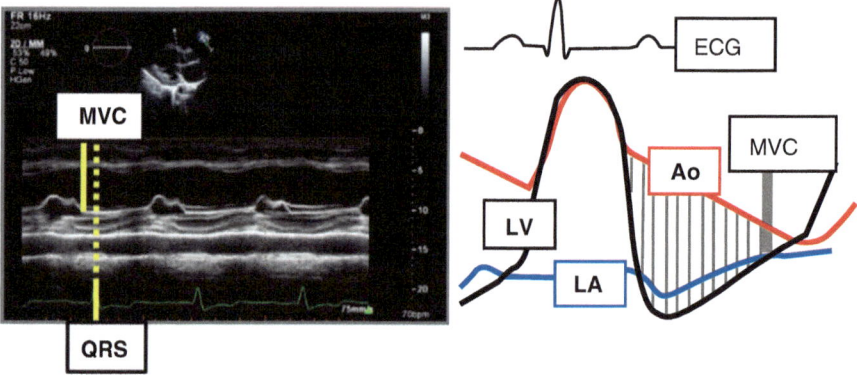

Fig. 10.7 Premature closure of the mitral valve. In acute severe AI, left ventricular (LV) diastolic pressure rises rapidly due to the poor compliance of the LV. The LV diastolic pressure can even rise above left atrial (LA) pressure, causing premature mitral valve closure (MVC). This is best identified on the mitral valve m-mode (left panel); the mitral valve leaflets closure can be seen preceding the QRS complex (*yellow lines*, left panel)

> M-mode tracings are particularly useful for accurate event timing; in the case shown it is evident that mitral valve closes even before the onset of the QRS confirming the highly elevated LVEDP.
>
> Occasionally, fine oscillations can be seen on the ventricular aspect of the anterior mitral leaflet; while these are caused by AI jet hitting the anterior mitral leaflet, they **do not** correlate with severity or acuity of the AI; the hemodynamically important parameter to assess on the mitral valve m-mode is the **timing** of the valve closure.

- Mitral valve color Doppler—presence of diastolic MR.

    As mentioned, an acute rise in LV diastolic pressure can cause the LV pressure to exceed the left atrial pressure and result in premature closure of the mitral valve.

    This mitral valve closure may be incomplete; either due to concomitant presence of incompetent mitral valve or due to the absence of the normal mechanisms responsible for complete and effective sealing of the mitral valve (i.e. decrease in atrial pressure during atrial relaxation and rapid LV pressure rise).

    Since the LV pressure at this point is higher than the LA pressure, diastolic mitral regurgitation may be seen.

    Similar to other instances of diastolic MR (see Chap. 8), the pressure gradient driving this flow is lower than the typical systolic MR driving gradient, and the resultant diastolic MR is of low velocity.

    Important to understand that the clinical significance of diastolic MR in the setting of acute AI is that it serves as a marker for the rapid elevation in the LV diastolic pressure which is the result of the AI occurring on a noncompliant ventricle; typically, the hemodynamic significance of the diastolic MR itself is negligible.

10.3 Summary and Final Points

- Regardless of the cause of aortic insufficiency (e.g., valvular, aortic root pathology, etc.), accurate quantification is essential for prognostication and treatment planning.
- Special attention should be paid to parameters that suggest hemodynamically significant AI with increased LV end-diastolic pressure; these may be signs of impending hemodynamic collapse and may necessitate prompt intervention.
- Although typically hemodynamically significant AI is also severe AI, there can be physiologic reasons for hemodynamically significant, non-severe AI (i.e., acute AI), or technical reasons, which may mask the echocardiographic appearance of hemodynamically significant AI.
- Occasionally echocardiographic findings may precede the clinical deterioration; proper consultation and timely intervention may be required to prevent hemodynamic collapse.

There's a Hole in the Heart: Part I

Abstract

Intracardiac shunts can occur at various locations and may be the result of a congenital heart defect—either as an isolated defect or as part of a complex congenital anomaly, or the result of other intracardiac pathologies like an infectious disease. Full evaluation is critical in order to offer optimal therapy—intervene when needed, or avoid intervention that can have detrimental effects. When investigating an atrial septal defect several parameters must be analyzed in order to obtain a comprehensive evaluation. These include type and size of the defect, magnitude and direction of the shunt, cardiac response to the overload (chamber size and function), and pulmonary pressures and pulmonary vascular resistance.

Keywords

Atrial septal defect · ASD · Shunt flow · Qp: Qs · Eisenmenger syndrome · Pulmonary hypertension · Pulmonary vascular resistance

Introduction

- Intracardiac shunts can occur at various locations (atrial, ventricular, arterial, and more).
- They can be the result of a congenital heart defect—either as an isolated defect or as part of a complex congenital anomaly, however, they can also be the result of other intracardiac pathology like an infectious disease.
- In this chapter, we will interrogate several case studies that will demonstrate what hemodynamic information can be derived from evaluation of intracardiac shunt flows.
- A full evaluation is critical in order to offer optimal therapy—intervene on time when needed, or avoid intervention that can have detrimental effects.

11.1 Approach to ASD Evaluation

- When an atrial septal defect is found, several questions need to be answered in order to decide on the best course of action.
  - Anatomic type of ASD.
  - Shunt direction.
  - Magnitude of the interatrial shunt.
  - Effect of the shunt on chamber size and function.
  - Presence of pulmonary hypertension.
  - Estimation of the pulmonary vascular resistance.

11.2 Case Presentation

- A 41-year-old woman presented to the emergency department with transient right-hand weakness and speech impairment; symptoms lasted approximately a minute and resolved spontaneously.
- Past medical history was unrevealing.
- Head CT did not show any acute findings; Carotid Doppler was normal.
- TTE followed by TEE were performed as part of the workup for embolic source; large secundum atrial septal defect with left to right shunt was seen (Fig. 11.1).

11.2.1 Anatomic Type of ASD

- The most common type of ASD is the secundum ASD; it is also the only anatomical type that is potentially amenable to percutaneous closure.
- Other types include: primum, sinus venosus (SVC/IVC type), and coronary sinus ASD.
- For a detailed evaluation of the various types of ASD please refer to other sources.

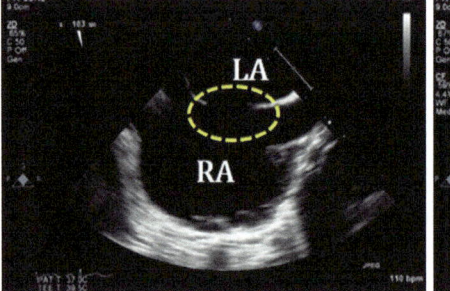 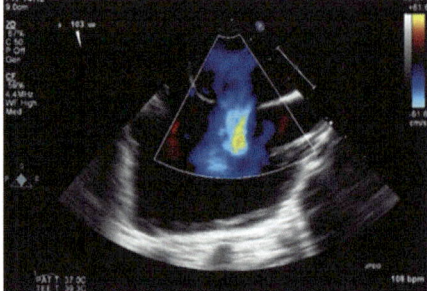

Fig. 11.1 Secundum atrial septal defect (ASD). TEE bicaval view showing a large secundum ASD (left panel, *yellow oval*). Color Doppler image from the same view (right panel) shows left to right flow through the ASD (LA—left atrium, RA—right atrium)

11.2.2 Shunt Direction

- Shunt flow is determined by the "downstream" pressure; meaning flow across the ASD is passive, down a pressure gradient.
- Left and right atrial pressures are very similar and hence they are not the main determinants of the shunt direction.
- The pressure "downstream," i.e., the pulmonary arterial pressure is the main determinant of shunt direction.
- As long as pulmonary pressures are low (and in particular pulmonary vascular resistance is low), the flow across the ASD is left to right.
- If the pulmonary vascular resistance increases and pulmonary hypertension develops, the left to right flow can decrease, and ultimately may even reverse.
- Color Doppler is helpful in assessing shunt direction by direct visualization of the flow across the ASD; on TTE, subcostal view is particularly helpful as the flow is nearly parallel to the US beam direction; on TEE, the bicaval view often allows for clear visualization of the interatrial flow.
- Spectral tracing of the shunt flow is also very helpful in delineating the shunt direction (Fig. 11.2); using a slow sweep speed provides a longer assessment of the flow, such that the effect of the respiratory cycle on the shunt flow can be assessed.
- In the case presented, clear left to right shunt was visualized both by color Doppler and spectral Doppler.

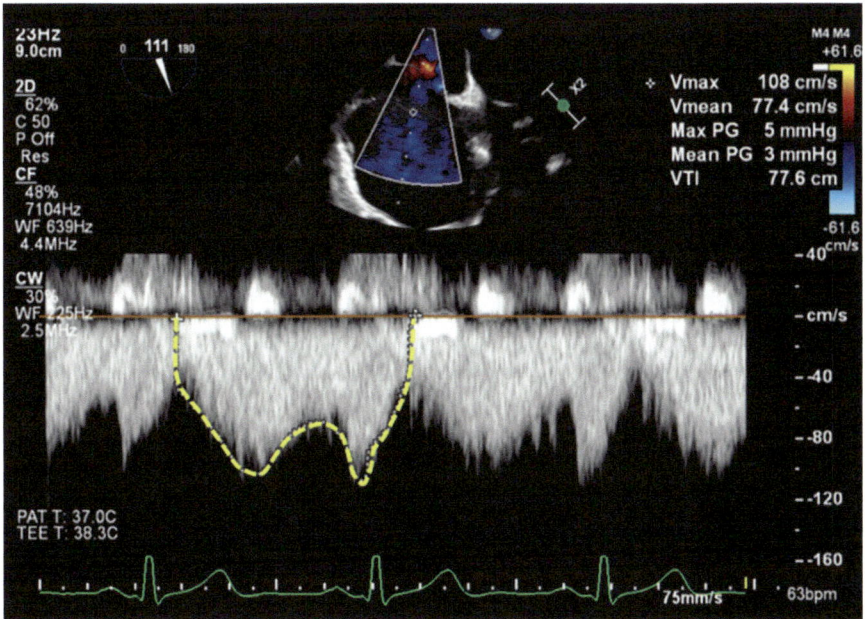

Fig. 11.2 ASD flow spectral Doppler. Continuous-wave Doppler of ASD flow obtained from TEE bicaval view. The flow is continuous, throughout systole and diastole, and low velocity

11.2.3 Magnitude of the Interatrial Shunt

- The magnitude of the interatrial shunt refers to the volume of blood that flows across the atrial septal defect.
- Two factors determine the magnitude of the interatrial shunt:
  - Defect size
  - Pulmonary pressure/pulmonary vascular resistance
- While large shunt flow typically means a large defect size, the reverse is not always the case; low shunt flow may be the result of a small ASD, however, may also be the result of elevated downstream pulmonary pressures.
- There are several ways by which to estimate the shunt flow on echocardiography.

Qp:Qs Calculation
- Qp: Qs calculation means comparing the flow across the pulmonary circulation (Qp) to the flow across the systemic circulation (Qs).
- Under normal circumstances (no shunt present) Qp:Qs ratio is 1:1; the pulmonary circulation cardiac output equals the systemic circulation cardiac output.
- When there is an intracardiac shunt, these outputs are unequal; in the presence of left to right shunt, the pulmonary circulation flow is higher than that of the systemic circulation.
- As discussed in previous chapters, stroke volume can be calculated at any area of the heart where cross-sectional area can be measured and spectral Doppler can be acquired.
- For purposes of Qp:Qs calculation, the easiest is to compare the flow across the RVOT and LVOT (Fig. 11.3).

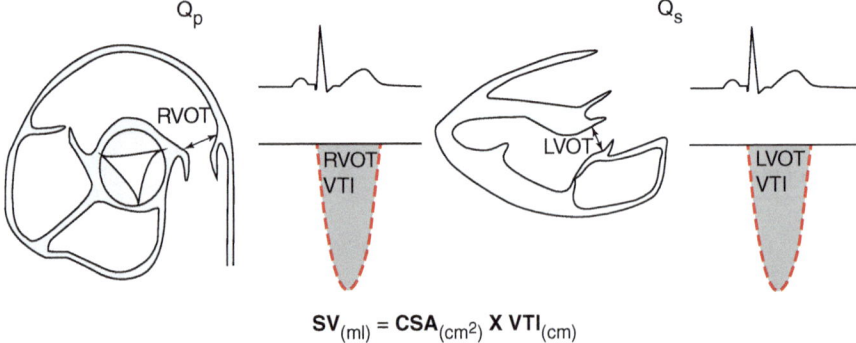

$$SV_{(ml)} = CSA_{(cm^2)} \times VTI_{(cm)}$$

Fig. 11.3 Calculating Qp:Qs. Comparing the flow across the pulmonary circulation (Qp) with the flow across the systemic circulation (Qs) can provide a quantitative assessment of the magnitude of intracardiac shunt. Typically RVOT diameter and VTI can be measured on the RVOT outflow view or the parasternal short-axis view at the level of the aortic valve. The LVOT diameter is typically measured on the PLAX view; LVOT VTI is acquired from apical five-chamber view or apical long-axis view. Flow volume is calculated as Cross-Sectional Area x VTI and the ratio between the two calculated volumes is determined

11.2 Case Presentation

- Typically RVOT diameter and VTI can be measured on the RVOT outflow view or the parasternal short-axis view at the level of the aortic valve.
- LVOT diameter is typically measured on the PLAX view; LVOT VTI is acquired from apical five-chamber view or apical long-axis view.
- $Qp = \pi \left(\dfrac{RVOT_d}{2}\right)^2 \times RVOT_{VTI}.$
- $Qs = \pi \left(\dfrac{LVOT_d}{2}\right)^2 \times LVOT_{VTI}.$
- Comparing the two gives the Qp:Qs ratio.
- There are no clear cutoffs defining when a closure procedure is indicated; generally speaking, Qp:Qs > 1.5: is considered significant; however, it is important to take into consideration all available parameters when deciding about a possible closure.
- It should be noted that there could be technical difficulties limiting the accuracy of Qp:Qs calculation; the main one is related to inaccuracy in measuring RVOT/LVOT diameter. Inaccurate measurement of the diameter introduces a significant error to the calculation since the measurement is squared when calculating cross-sectional area.
- Optimization of the LVOT/RVOT visualization is extremely important in order to enhance the reliability of the calculation.
- In the case presented, the following data were obtained:
    - RVOT
        Diameter—3.2 cm
        VTI—35 cm
        Stroke volume—281 cc
    - LVOT
        Diameter—2.3 cm
        VTI—26 cm
        Stroke volume—108 cc
- **Qp:Qs** thus calculated to be 2.6:1.

Shunt Flow Measurement
- The flow across the ASD can occasionally be directly measured.
- As discussed above, flow volume can be calculated at any location where cross-sectional area can be measured and spectral Doppler can be obtained.
- If the CSA of the ASD is known and spectral Doppler across the ASD can be obtained, the shunt volume can be calculated as: Shunt Volume $= \pi \left(\dfrac{ASD_d}{2}\right)^2 \times ASD_{VTI}.$
- In the case presented: ASD diameter ~ 1.6 cm, and the ASD VTI = 77 cm ➜ ASD shunt flow = 155 ml.
- To double check if this calculation makes sense, shunt volume and the LVOT stroke volume can be used to calculate the Qp.

- Qp = Qs + Shunt Volume.
- Using the numbers from the case presented:
- Qs = 108 cc, Shunt volume = 155 cc, Qp = 108 + 155 = 263 cc ➜ Qp:Qs = 2.4:1, numbers very similar to those obtained by the previous method.
- Technical difficulties related to ASD measurement should also be pointed out:
  - The ASD may not be circular in shape; often it is more elliptical in shape. Using the biplane view (available on 3D TEE) can allow measurement of the ASD diameter in two orthogonal views.
  - Alternatively, if a good 3D image of the defect can be obtained, the CSA area can be directly measured by utilizing the 3D software and post-processing of the image.
  - The ASD size may change during the cardiac and inspiratory cycle; averaging several measurements or matching the timing of the anatomical measurement and the Doppler measurement should increase the accuracy of the calculation.
  - If the entire ASD cannot be visualized or it is a fenestrated ASD, CSA may be difficult to assess and may introduce an unacceptable error into the calculation.
- The information obtained by this technique should be evaluated in conjunction with all the other parameters by which the ASD is assessed.

11.2.4 Effect of the Shunt on Chamber Size and Function

- A significant interatrial shunt results in volume overload of the right-sided chambers.
- Right ventricular and right atrial dilatation are commonly seen with clinically significant ASD.
- Increased RVOT/pulmonary artery flow can be seen (without the presence of pulmonary stenosis).
- Pulmonary artery dilatation may be seen as well.
- Abnormal interventricular septal motion can be seen with abnormal eccentricity index.
- In advanced cases, RV failure can occur with decreased RV systolic function; TAPSE, tissue Doppler RV S', contractile function, etc. should all be assessed.

11.2.5 Presence of Pulmonary Hypertension

- Identifying the presence and severity of pulmonary hypertension in patients with ASD is extremely important for proper treatment decisions.
- Pulmonary pressures can be estimated using the techniques discussed in Chaps. 3 and 4; most commonly, the TR peak velocity is used to calculate the pulmonary artery systolic pressure.

- In the case presented, peak TR velocity measured 3.9 m/sec, IVC was normal in size and collapsed with inspiration → pulmonary artery systolic pressure estimated ~63 mmHg.
- Pulmonary pressure can be elevated due to either one of two factors:
  - Elevated flow across the pulmonary circulation.
  - Elevated resistance across the pulmonary circulation.
- It is of paramount importance to distinguish between these two causes; elevated pressures due to high flow will likely resolve upon correction of the ASD and should not be considered a contraindication to ASD closure. However, elevated pulmonary vascular resistance may be a contraindication to ASD closure as pulmonary pressure can remain elevated even after ASD closure and RV failure may ensue.

11.2.6 Estimation of Pulmonary Vascular Resistance

- Pulmonary vascular resistance can be estimated using the relationship between pressures, flow, and resistance:
- $\text{Flow} = \dfrac{\text{Driving Pressure}}{\text{Resistance}}$.
- Rearranging the formula allows calculating the resistance:
- $\text{Resistance} = \dfrac{\text{Driving Pressure}}{\text{Flow}}$.
- For calculating the PVR, the driving pressure is the pressure drop across the pulmonary circulation: $\Delta P = mean\ PAP - $ Left Atrial Pressure (LAP).
- The flow across the pulmonary circulation is the Qp; note that in a case with intracardiac shunt, the **correct** cardiac output must be plugged into the equation as the systemic and pulmonary flows are **not equal.**
- In the case presented, the PI end-diastolic velocity measured 1.7 m/sec → PI diastolic pressure estimated to be 15 mmHg.
- $\text{Mean PAP} = \dfrac{1}{3}\text{sysPAP} + \dfrac{2}{3}\text{diasPAP} = 21 + 10 = 31\,\text{mmHg}$.
- The patient was in no respiratory distress with normal left atrial size → reasonable to assume normal LAP, approximately 10 mmHg.
- Patient's heart rate during the exam was approximately 80 bpm; calculated pulmonary system cardiac output $270 \times 80 \approx 22{,}000 = 22$ lit/min.
- Using these numbers:
- $\text{PVR} = \dfrac{\Delta P}{\text{CO}} = \dfrac{21}{22} = 0.9\,\text{Woods Units}$.
- Despite the elevated pulmonary pressure, the PVR is low in this patient; given the magnitude of the calculated shunt and the elevated pulmonary blood flow, it is not surprising that PVR is low; in fact, in order to be able to pass 22 lit every minute across the pulmonary circulation it is almost expected that the PVR is normal or even low.

11.3 Case Presentation

- A 57-year-old woman presented to the emergency department with increasing shortness of breath and fatigue.
- Her symptoms started sub-acutely, over the course of the several months preceding her admission.
- She described increasing dyspnea with significantly decreasing effort tolerance.
- Her echocardiogram showed a large secundum ASD (Fig. 11.4).
- As discussed above, several questions need to be answered when assessing an atrial septal defect.
    - Anatomic type of ASD (see above—refer to other sources for full discussion).
    - Shunt direction.
    - Magnitude of the interatrial shunt.
    - Effect of the shunt on chamber size and function.
    - Presence of pulmonary hypertension.
    - Estimation of the pulmonary vascular resistance.

11.3.1 Shunt Direction

- Shunt flow is determined by the pressure difference between systemic and pulmonary circulations; as long as pulmonary pressures/pulmonary vascular resistance are low, flow across an ASD is left to right.
- Additionally, when pulmonary pressures are low, the degree of shunting is proportional to the size of the ASD; the larger the ASD the larger the shunt.

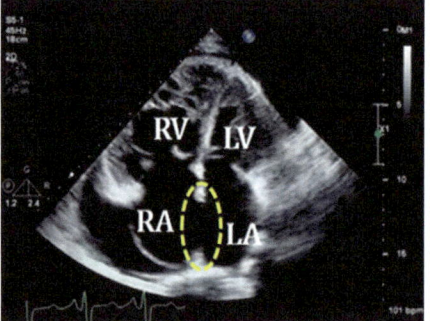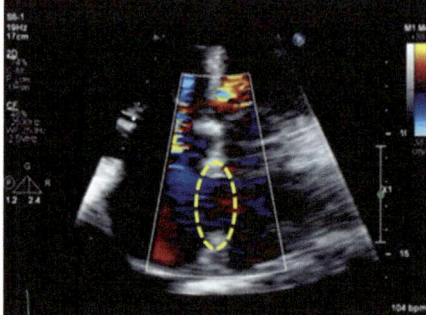

Fig. 11.4 Large secundum atrial septal defect (ASD) with Eisenmenger syndrome. Apical four-chamber view showing a large defect (*yellow oval*, left panel) in the interatrial septum, consistent with secundum ASD. Severe right ventricular and right atrial dilatation seen as well. Color Doppler image from the same view (right panel) shows non-specific, low-velocity color pattern; shunt direction cannot be adequately discerned on this image (LA—left atrium, LV—left ventricle, RA—right atrium, RV—right ventricle)

11.3 Case Presentation

- However, when pulmonary pressures and vascular resistance increase, left to right shunt decreases, and in extreme situation may even reverse and become right to left (Eisenmenger syndrome).
- When left to right shunting decreases, it may become difficult to visualize the shunt by color Doppler; spectral tracing may also become non-specific as the typical continuous left to right flow is lost.
- It is important to understand that lack of visualization of color flow across an atrial septal defect does **not** imply a clinically insignificant disease.

11.3.2 Magnitude of the Interatrial Shunt

- The magnitude of the shunt refers to the volume of blood that is shunted across the defect.
- As mentioned earlier, it is determined by the defect size as well as by the "downstream" pressures / resistance in the pulmonary vs. the systemic circulation.
- The techniques used by echocardiography to quantify the shunt include:
    - Calculating the Qp:Qs (ratio of pulmonary flow to systemic flow).
    - Calculating the volume that flows across the septal defect (using measurement of the size of the defect and spectral Doppler of the flow at the defect).
- For the patient presented, the following data were obtained:
    - **RVOT**
        Diameter—3 cm
        VTI—9 cm
        Stroke volume—63 cc
    - **LVOT**
        Diameter—1.9 cm
        VTI—18 cm
        Stroke volume—51 cc
- Qp:Qs thus calculated to be 1.2:1.
- The ASD diameter was measured approximately 1.8 cm; the spectral Doppler tracing of the flow across the ASD was very faint and could not be reliably traced → direct calculation of the shunt flow could not be performed.

11.3.3 Effect of the Shunt on Chamber Size and Function

- As mentioned, a significant interatrial shunt results in volume overload of the right-sided chambers; right ventricular and right atrial dilatation are commonly seen.
- The patient had severe RV/RA dilatation, PA dilation, and her IVC was dilated and plethoric.
- Notably, signs of RV pressure overload were also seen; parasternal short axis view was significant for septal flattening ("D shaped" septum) with increased eccentricity index both during diastole and systole (Fig. 11.5).
- In addition, RV function was markedly depressed.

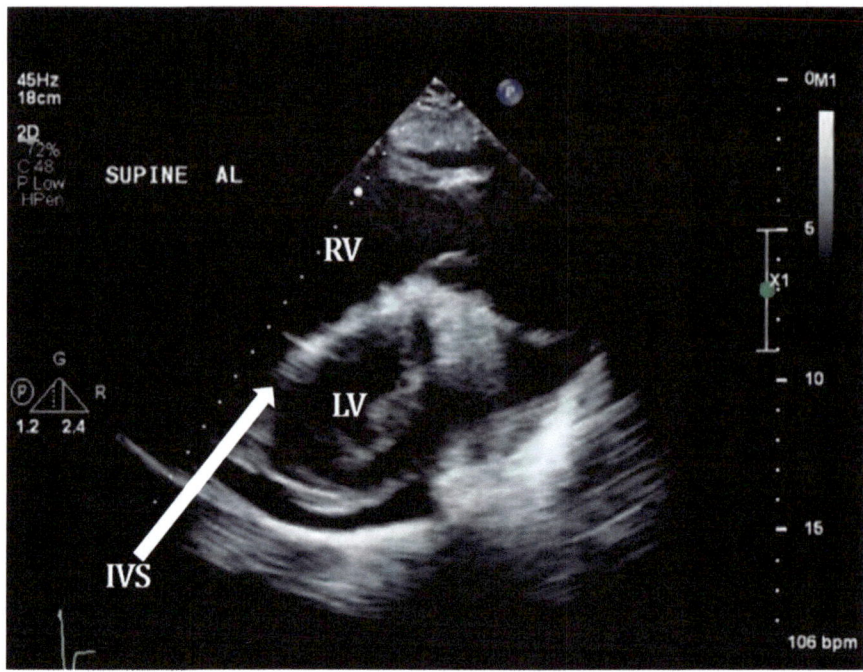

Fig. 11.5 D-shaped septum in severe pulmonary hypertension. Systolic frame from parasternal short-axis view showing a markedly dilated right ventricle (RV). In addition, the interventricular septum (IVS) is flattened, such that left ventricular (LV) anteroposterior diameter is larger than its septal–lateral diameter (increased eccentricity index). As this is observed during systole it implies significant pressure overload of the right ventricle

11.3.4 Presence of Pulmonary Hypertension

- The patient had moderate TR; peak TR velocity 3.7 m/sec → TR peak gradient 55 mmHg (Fig. 11.6).
- Mild PI was noted as well with end-diastolic velocity of 2 m/sec → end-diastolic PI gradient 16 mmHg.
- The IVC was dilated and plethoric, indicating elevated right atrial pressure.
- From the above numbers:
  - PA systolic pressure—70 mmHg
  - PA diastolic pressure—30 mmHg
  - Mean pulmonary arterial pressure—43 mmHg
- These numbers are suggestive of severe pulmonary hypertension

11.3.5 Estimation of Pulmonary Vascular Resistance

- Knowing the pulmonary vascular resistance is essential for proper treatment decisions.

11.3 Case Presentation

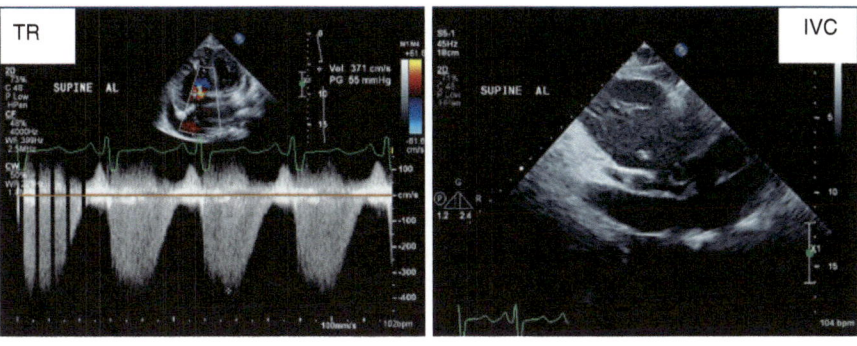

TR Peak V = 3.7m/sec
ΔP = 55mmHg

Dilated and plethoric IVC →
elevated RAP

$$PA_{sys}\ P = \Delta P_{RV\text{-}RA} + RAP \approx 70\,mmHg$$

Fig. 11.6 Pulmonary artery pressure calculation. Images from the echocardiogram of the second patient presented; Tricuspid regurgitation (TR) peak velocity was 3.7 m/sec; IVC was dilated and plethoric indicating elevated right atrial pressure. The pulmonary artery systolic pressure was estimated at approximately 70 mmHg

- Presence of markedly elevated PVR is a contraindication to ASD closure as pulmonary hypertension may be irreversible; closure of the ASD may lead to right heart failure and collapse.
- Summarizing the available data for the patient:
  - **Systemic SV**—51 cc
  - **Pulmonary SV**—63 cc
  - **Qp:Qs**—1.2:1
  - **TR Velocity**—3.7 m/sec
  - **PI EDV**—2 m/sec
  - **RAP estimate**—15 mmHg
  - **PA systolic pressure**—70 mmHg
  - **PA diastolic pressure**—30 mmHg
  - **Mean PA pressure**—43 mmHg
- The patient's heart rate during the examination was approximately 100/min.
- There were no signs of pulmonary congestion on chest X-ray or on examination → reasonable to assume relatively normal left atrial pressure, ~10 mmHg.
- Using the above data the following can be calculated:

$$\text{Transpulmonary Pressure Drop}: 43 - 10 = 33\,mmHg$$

$$\text{Pulmonary flow}\,(\text{cardiac output}): 63 \times 100 = 6.3\,\text{lit}/\min$$

$$PVR = \frac{\text{Transpulmonary }\Delta P}{\text{Pulmonary CO}} = \frac{33}{6.3} = 5.2\,\text{Woods Units}\,(\text{markedly elevated})$$

11.4 Summary and Final Points

- When analyzing intracardiac shunts it is essential to assess all available information in order to understand the effect of the shunt on the heart.
- Parameters to pay attention to include the size of the defect, the magnitude of the shunt, the cardiac response to the overload (chamber size and function), pulmonary pressures, and pulmonary vascular resistance.
- Comparing the two patients presented in this chapter, it can be seen that despite the ASD size being very similar and calculated pulmonary artery systolic pressures being comparable, these two patients were vastly different.
- While the first patient will likely benefit from ASD closure (both for prevention of future RV failure and prevention of repeat paradoxical embolic events), the second patient requires further assessment before decisions can be made.
- Findings suggestive of severe pulmonary hypertension, elevated PVR and near reversal of the interatrial shunt require confirmation with right heart catheterization and direct measurements of pressures and resistance.
- Full echocardiographic evaluation of all available parameters is critical for comprehensive understanding of the impact of ASD on the heart.
- Only when evaluating all these parameters (and paying attention to possible pitfalls) a true understanding of the defect and its hemodynamic significance can be gained.

There's a Hole in the Heart: Part II

12

Abstract

Intracardiac shunts can occur at various locations and may be the result of a congenital heart defect, other intracardiac pathology (i.e., infectious process), or iatrogenic complications of surgical or structural interventions. When analyzing a ventricular septal defect several parameters must be assessed in order to obtain a comprehensive evaluation. These include type and size of the defect, magnitude and direction of the shunt, cardiac response to the overload (chamber size and function), pulmonary pressures, and pulmonary vascular resistance. Interrogating VSD flow can provide important hemodynamic information and data regarding intracardiac pressures. It is important to make sure that all the flows present are interrogated and measured and obtained parameters should "make sense" when evaluated together; discrepancy in data obtained should prompt a thorough evaluation in order to attain an understanding of the nature of the discordant data.

Keywords

Ventricular septal defect · Qp:Qs · Pulmonary hypertension · Pulmonary vascular resistance · Double chamber RV · Subpulmonic stenosis

Introduction

- Intracardiac shunt can be found at the ventricular level, in the interventricular septum.
- Ventricular septal defect (VSD) can be caused by:
  - Congenital anomaly—either isolated defect or as part of complex congenital syndrome.
  - Complication of various cardiovascular pathologies (e.g., myocardial infarction and endocarditis).

- Iatrogenic complication from surgical or interventional procedures.
- Regardless of etiology, a comprehensive assessment of the VSD can provide important insight into the underlying pathology and the hemodynamic status.
- In this chapter, we will review several case studies to illustrate how to interrogate a VSD and what can be deduced from such evaluation.

12.1 Approach to VSD Evaluation

- Similar to assessment of ASD, several parameters need to be assessed in order to have a complete understanding of the VSD and its effect on the heart.
- Parameters to assess include:
  - Anatomic type of VSD.
  - Shunt direction and magnitude of the interventricular shunt.
  - Effect of the shunt on chamber size and function.
  - Presence of pulmonary hypertension and PVR assessment.

12.2 Case Presentation

- A 63-year-old man presented to the emergency department with a late presentation of anterior wall myocardial infarction.
- Medical treatment was initiated and the patient was admitted to the CCU.
- Notably, the patient was dyspneic and appeared volume overloaded on physical examination → diuretic therapy and noninvasive ventilation started.
- Echocardiogram was performed and showed large rupture at the apical interventricular septum with significant left to right shunt (Fig. 12.1).

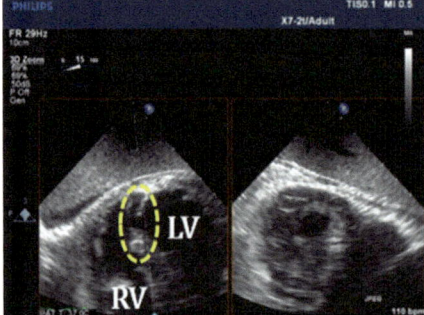
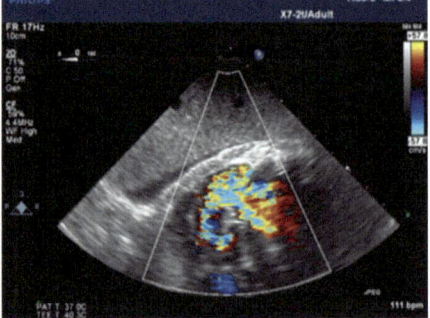

Fig. 12.1 Post-infarct ventricular septal defect. TEE deep transgastric view showing a large tear at the apical interventricular septum (left panel, *yellow oval*). Color Doppler image from the same view (right panel) showing large turbulent left-to-right flow across the septal tear

12.2 Case Presentation

12.2.1 Anatomic Type of VSD

- The interventricular septum is made up of several components: membranous, trabecular/muscular, inlet, outlet, and the ventriculoatrial septum.
- Congenital VSD can involve any part of the septum; different congenital syndromes typically involve specific parts of the septum.
- For more detailed discussion of the various anatomic types, please refer to other texts.
- Acquired VSDs occur in the area associated with the pathology that caused them; for instance, VSD complicating surgical myectomy is typically in the membranous septum.
- In the case presented, the patient had an anterior wall MI due to total occlusion of the left anterior descending (LAD) artery; the VSD was seen at the apical septum (typically supplied by the LAD).

12.2.2 Quantifying the Degree of Shunt

- For the patient presented above, the anatomic and color Doppler assessment showed that there was a large tear in the interventricular septum with significant left to right flow.
- Quantifying the shunt is an important part of the assessment of any VSD.
- Quantification of VSD shunt can follow the same techniques as described for ASD; Qp:Qs ratio can be calculated and direct measurement of flow across the VSD can be attempted.
- For the patient presented above, the following data were obtained (Figs. 12.2 and 12.3):

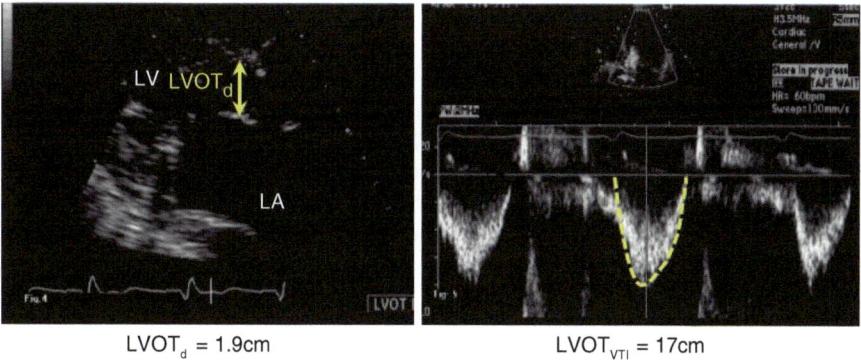

$LVOT_d = 1.9cm$ $LVOT_{VTI} = 17cm$

$SV_{Left} = CSA \times VTI = 2.8 \times 17 = 48cc$

Fig. 12.2 Qp:Qs calculation—left sided stroke volume (SV_{left}). The LVOT diameter ($LVOT_d$) is measured from the parasternal long-axis view; the $LVOT_{VTI}$ is measured from an apical view. SV_{left} is calculated as LVOT Cross-sectional area (LVOT CSA) x $LVOT_{VTI}$ (LA—left atrium, LV—left ventricle)

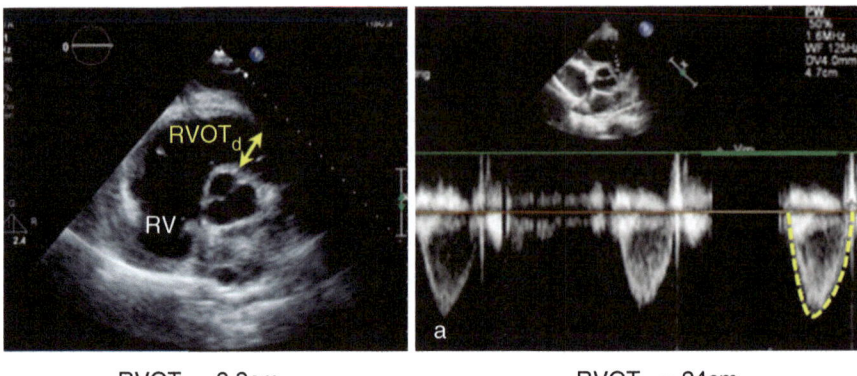

$RVOT_d = 2.2 cm$ $RVOT_{VTI} = 24 cm$

$SV_{right} = CSA \times VTI = 3.8 \times 27 = 91 cc$

Fig. 12.3 Qp:Qs calculation—right-sided stroke volume (SV_{right}). The RVOT diameter ($RVOT_d$) is measured from the parasternal short-axis view at the aortic valve level; the $RVOT_{VTI}$ is measured from the same view. SV_{right} is calculated as RVOT Cross-sectional area (RVOT CSA) x $RVOT_{VTI}$ (RV—right ventricle)

	RVOT	LVOT
Diameter	2.2 cm	1.9 cm
VTI	24 cm	17 cm
Stroke volume	91 cc	48 cc

- Thus, Qp:Qs calculated to be 1.9:1.
- Direct measurement of the shunt could not be performed in this case; often post MI VSDs have complicated anatomy with multiple tracks and tears at the injured area of the septum, making accurate diameter measurement impossible.

12.2.3 VSD Flow and Intracardiac Pressures

- Most often, a VSD is a communication between the left and right ventricle.
  - The exception is a Garbode defect; a ventriculo-atrial communication connecting the LV and the RA; either directly (via a defect in the ventriculo-atrial septum) or a VSD with a perforation of tricuspid valve leaflet.
- The driving force for the flow across the VSD is the pressure difference between the left ventricle and the right ventricle.
- When intracardiac pressures are relatively normal (Fig. 12.4):
  - LV pressure is typically approximately 120/10.
  - RV pressure is approximately 20/5.
  - LV-RV pressure difference (ΔP_{LV-RV}):
    During systole $\Delta P_{LV-RV} \approx 100$ mmHg.
    During diastole $\Delta P_{LV-RV} \approx 0$–5 mmHg.

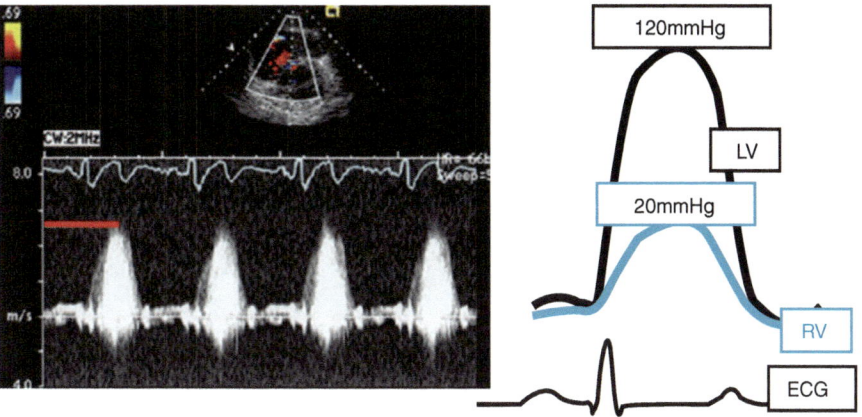

Fig. 12.4 Typical VSD flow. The driving force for the flow across the VSD is the pressure difference between the left ventricle (LV) and the right ventricle (RV). When intracardiac pressures are relatively normal, the systolic pressure difference is typically high, giving rise to high-velocity systolic flow; during diastole LV and RV pressures are nearly equal, such that diastolic VSD flow is minimal

- These pressure differences make the classic VSD flow a high velocity (~5 m/sec), systolic only flow.
- In order to accurately assess the flow characteristics across the VSD it is important to interrogate the flow from multiple views, to assure the US beam is as parallel as possible to the VSD flow (such that the true highest velocity is measured).
- In the case presented the peak velocity across the VSD was 3.9 m/sec (Fig. 12.5) → ΔP_{LV-RV} = 61 mmHg (lower than typical VSD gradient).
- The patient's blood pressure at the time of the echocardiogram was 125/64.
- Knowing the blood pressure and the peak VSD velocity / gradient allows calculation of the RV systolic pressure (assuming LV systolic pressure = aortic systolic pressure).
- In the case presented: RV systolic pressure = 125−61 = 64 mmHg.
- RV systolic pressure typically equals the pulmonary artery systolic pressure; it can be concluded that this patient had elevated pulmonary artery systolic pressure.
- This finding can be confirmed by interrogating the TR velocity (Fig. 12.6):
  - Peak TR velocity—3.8 m/sec.
  - IVC—normal in size with normal inspiratory collapse.
  - Thus, RV systolic pressure estimated to be ≈ 58 + 3 = 60–65 mmHg.
- The patient also had mild pulmonic insufficiency; interrogating the PI spectral Doppler provides an estimate of the pulmonary artery diastolic pressure (see Chap. 4):
  - PI EDV—1.2 m/sec.
  - Thus, PA diastolic pressure estimated to be ≈ 6 + 3 = 8–10 mmHg.

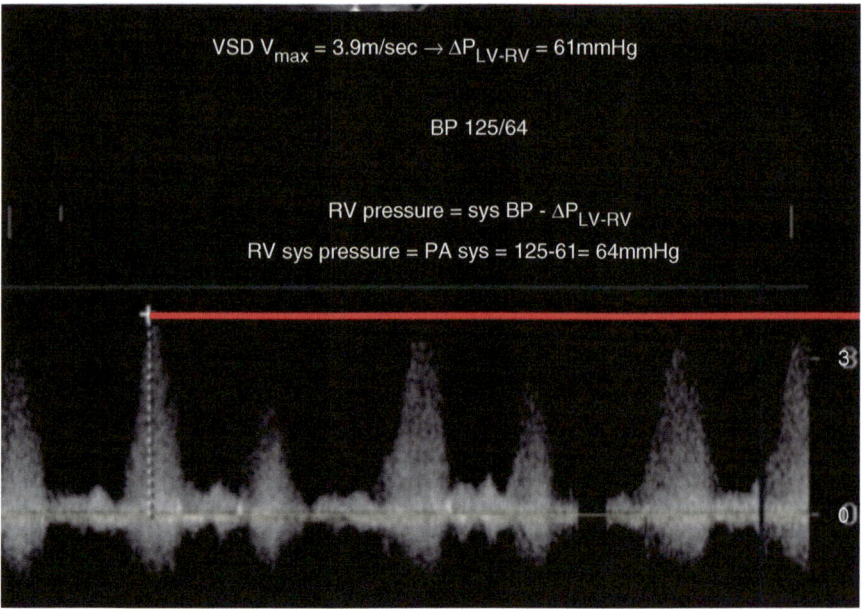

Fig. 12.5 VSD spectral tracing (post-infarct VSD). The peak VSD velocity is low—3.9 m/sec. This correlates with peak LV-RV gradient (ΔP_{LV-RV}) of 61 mmHg. Blood pressure at the time of the echo: 125/64. The patient had no aortic stenosis or any other reason for gradient between the left ventricle (LV) and the aorta → LV systolic pressure can be assumed to be 125 mmHg. The RV systolic pressure, and hence the pulmonary artery systolic pressure can thus be calculated as: systolic blood pressure—ΔP_{LV-RV} = 64 mmHg (high pulmonary artery systolic pressure)

- From the estimated systolic and diastolic pulmonary artery pressures, mean PA pressure can be calculated: $\text{mean PAP} = \frac{1}{3}\text{sysPAP} + \frac{2}{3}\text{diasPAP} = 27\,\text{mHg}$.
- Knowing the mean pulmonary artery pressure and the flow across the pulmonary circulation allows calculation of the pulmonary vascular resistance.
- In the case presented:
  - Mean PAP estimated 27 mmHg
  - The patient was dyspneic, requiring diuretics and noninvasive ventilation → his left atrial pressure (LAP) was clearly elevated. While an exact measurement of his LAP could not be obtained by echo, it is reasonable to assume LAP of at least 15 mmHg.
  - The calculated right sided stroke volume was 91 cc.
  - Heart rate at the time of the echo cardiogram—75/min.
  - Pulmonary flow was thus 91 × 75 = 6.83 lit/min.

12.2 Case Presentation

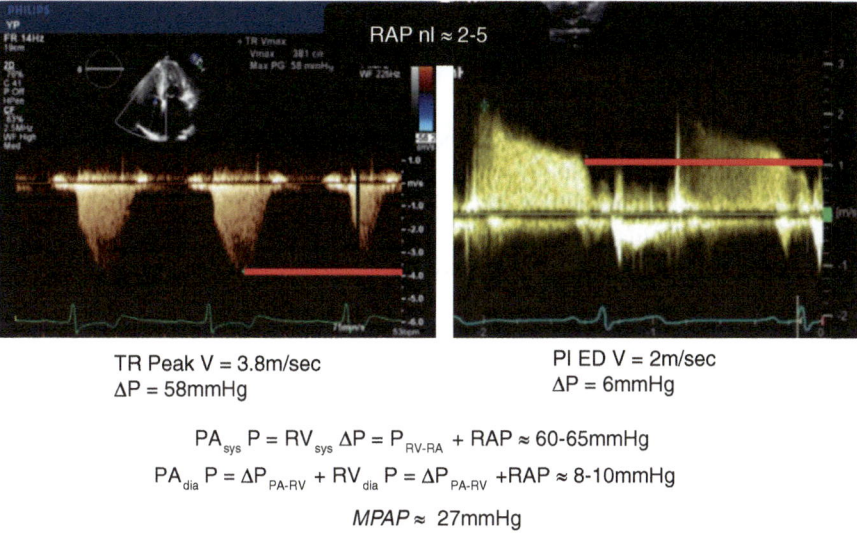

TR Peak V = 3.8m/sec
ΔP = 58mmHg

PI ED V = 2m/sec
ΔP = 6mmHg

$PA_{sys} P = RV_{sys} \Delta P = P_{RV-RA} + RAP \approx 60\text{-}65 mmHg$

$PA_{dia} P = \Delta P_{PA-RV} + RV_{dia} P = \Delta P_{PA-RV} + RAP \approx 8\text{-}10 mmHg$

MPAP ≈ 27mmHg

Fig. 12.6 TR/PI pulmonary pressure estimate (post-infarct VSD). The pulmonary artery pressure can also be calculated by the tricuspid regurgitation (TR) peak velocity and the pulmonic insufficiency (PI) end-diastolic velocity. TR peak velocity measured 3.8 m/sec, PI end-diastolic velocity—2 m/sec. Right atrial pressure (RAP) based on inferior vena cava size and collapsibility was normal. Using these data, the pulmonary artery pressure was calculated to be 65/10 mmHg

- $PVR = \dfrac{\text{Transpulmonary pressure drop}}{\text{Pulmonary cardiac output}} = \dfrac{27-15}{6.83} = 1.75 \, WU$.

12.2.4 Summary of Post MI VSD Case

- Patient admitted with late presentation of anterior wall MI.
- Course complicated by apical septal rupture creating a large VSD.
- Echocardiographic assessment showed:
  - Anatomic defect in the apical septum.
  - Significant color flow from left to right.
  - Shunt quantification with significant Qp:Qs (1.9:1).
  - Pulmonary hypertension (both by interrogating the VSD flow as well as the TR/PI flow).
  - Mildly elevated PVR.
- Treatment of post MI VSDs is very complicated; although surgical intervention offers the best survival benefit, it should be noted it is a very high-risk procedure with significant morbidity and mortality.

12.3 Case Presentation

- A 41-year-old man presented with increasing dyspnea and decreasing effort tolerance.
- Diagnosed with a "small hole in the heart" after a murmur was noted on physical examination in childhood.
- Echocardiogram showed (Fig. 12.7):
  - Biventricular dilatation with severe biventricular systolic dysfunction.
  - Membranous VSD, left-to-right flow, and unusual spectral Doppler pattern.

12.3.1 Anatomic Type of VSD

- Given the history of long standing murmur and a prior diagnosis of "hole in the heart" it appeared likely that the VSD that was seen was congenital and not acquired.
- In addition, there was no history of myocardial infarction, infection or any recent cardiac procedure that could have been complicated by a VSD.
- Based on the location of the VSD, as seen on the parasternal long- and short-axis views, the most likely variant was a congenital membranous VSD.
- Please refer to other texts for full description of the various types of VSDs.

12.3.2 Quantifying the Degree of Shunt

- As describe previously, the magnitude of the shunt is determined by the volume of blood the flows across the defect.
- This can be quantified either by calculating the Qp:Qs ratio or by direct measurement of the flow across the VSD (by measuring the VSD cross sectional area and the VTI of the flow as obtained by spectral Doppler).
- In this case, Qp:Qs ratio was calculated to be relatively low, approximately 1.2–1.3:1.

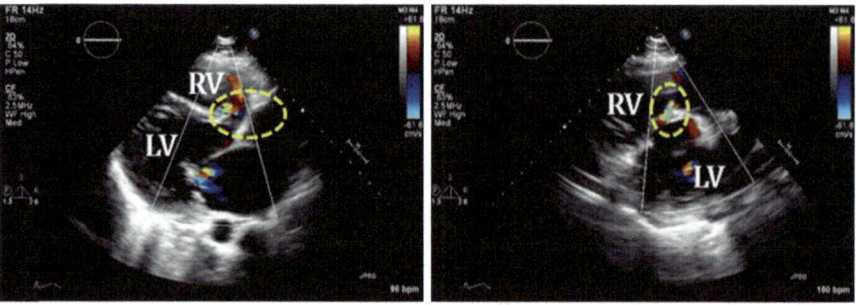

Fig. 12.7 Membranous VSD. Parasternal long-axis view (left panel) and parasternal short-axis view (right panel) showing a small membranous VSD (*yellow oval*) with left to right flow

12.3.3 VSD Flow and Intracardiac Pressures

- The spectral Doppler across the VSD (Fig. 12.8) was unusual on two counts:
  - Left-to-right flow was seen both during systole and diastole.
  - Peak systolic velocity across the VSD was low, only 3 m/sec.
- Diastolic flow across a VSD.
  - Presence of left-to-right flow during diastole signifies the presence of a pressure gradient between the LV and RV during diastole, which implies that LV diastolic pressure is elevated.
  - In the absence of mitral stenosis, LV diastolic pressure equals the left atrial pressure.
  - Thus, presence of left-to-right flow during diastole means that left atrial pressure is elevated.
  - Looking at the patient's spectral Doppler tracing, the diastolic VSD flow velocity is approximately 1.6 m/sec, which means that LV diastolic pressure is ~10 mmHg higher than RV diastolic pressure.
  - RV diastolic pressure equals the right atrial pressure; in this patient, IVC imaging showed a dilated and plethoric IVC, consistent with elevated RA pressure; typically RAP of 15 mmHg is assumed with these IVC parameters.
    Important to note though that RA pressure may be higher; it is by convention to use estimated RAP of 15 mmHg.
  - Assuming RAP of 15 mmHg → RV diastolic pressure is 15 mmHg.
  - With a VSD diastolic gradient of 10 mmHg, the estimated LV diastolic pressure and LA pressures are: 15 + 10 = 25 mmHg.
- Low peak systolic VSD velocity.

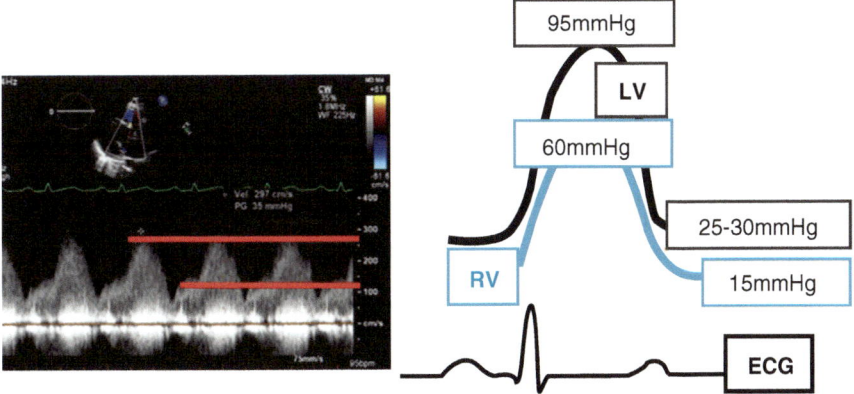

Fig. 12.8 Spectral Doppler and pressure tracings (VSD with cardiomyopathy). Spectral Doppler of the VSD flow (left panel) showing low-velocity systolic flow, as well as continuous flow during diastole. Low-velocity systolic flow implies a low systolic gradient between the left ventricle (LV) and the right ventricle (RV), which can be caused by low systolic blood pressure and high pulmonary artery/right ventricular systolic pressure. Continuous diastolic flow implies elevated LV diastolic pressure (meaning elevated left atrial pressure/wedge pressure)

- The driving force for the systolic VSD flow is the systolic pressure difference between the LV and RV (ΔP_{LV-RV}).
- As discussed above, when intracardiac pressures are within normal limits, there is a large pressure difference between LV and RV during systole; systolic VSD flow is thus typically a high-velocity flow.
- The peak velocity measured on the patient's spectral tracing was 3 m/sec, corresponding to a $\Delta P_{LV-RV} \approx 35$ mmHg.
- Low ΔP_{LV-RV} can be the result of low LV systolic pressure, high RV systolic pressure, or a combination of these two factors.
- The patient's blood pressure during the echocardiogram was 95/62; in the absence of any LV-aortic obstruction, LV systolic pressure equals aortic systolic pressure.
- Knowing LV systolic pressure allows calculation of the RV systolic pressure.
- *RV systolic pressure = LV systolic pressure − ΔP_{LV-RV} = 95 − 35 = 60 mmHg.*
- Typically, RV systolic pressure equals PA systolic pressure.
- Thus, from the measured VSD systolic peak velocity, the estimated patient's pulmonary artery systolic pressure is elevated.
- Summarizing the numbers obtained so far:

Qp:Qs	1.2:1
RAP	15 mmHg
VSD dias ΔP	10 mmHg
Estimated LAP	25 mmHg
BP	95/62 mmHg
VSD sys ΔP	35 mmHg
Sys PAP	60–65 mmHg

- The following day, the patient underwent left and right heart catheterization for further evaluation.
- Coronary angiogram was normal.
- LV angiogram showed severe diffuse LV dysfunction.
- Hemodynamics data:

LV (S/EDP)	90/34 mmHg
Ao (S/D/mean)	90/70/77 mmHg
Wedge	30 mmHg
PA (S/D)	65/30 mmHg
RV (S/EDP)	65/15 mmHg
RA (mean)	15 mmHg
HR	90 bpm

12.3.4 Summary of VSD with Cardiomyopathy Case

- Patient with a known congenital VSD presented with increasing dyspnea and decreasing effort tolerance.
- Shunt calculation revealed likely small left to right shunt.

12.4 Case Presentation

- Analysis of the VSD flow pattern was consistent with low left ventricular systolic pressure, high right ventricular systolic pressure, and elevated LV diastolic pressure.
- These echocardiographic findings were suggestive of low aortic pressure, elevated pulmonary artery systolic pressure, and elevated left atrial / wedge pressure.
- Findings on right heart catheterization confirmed precisely the hemodynamics as predicated by the echo evaluation.
- Most likely, the patient suffered from two unrelated disorders; a congenital small membranous VSD and later development of idiopathic dilated cardiomyopathy.
- Given the small shunt and the dilated right heart, it would be unlikely that the cardiomyopathy is the sequela of the VSD; for a VSD to cause severe pulmonary HTN and Eisenmenger syndrome it would classically be a large VSD rather than a restrictive, small VSD diagnosed by incidental murmur heard at a pediatric well-child examination.

12.4 Case Presentation

- A 48-year-old man was referred for echocardiogram for evaluation of VSD and pulmonary pressures.
- The patient was known to have a VSD, however, the pulmonary pressures were unclear on prior examinations.
- No precise data regarding the type of VSD or prior interventions could be accurately obtained from the patient.

12.4.1 Anatomic Type of VSD

- VSD was identified on the TTE with left to right shunt.
- Based on the location of the VSD, as seen on the parasternal long- and short-axis views, the most likely variant was a congenital supracristal VSD (Fig. 12.9).
- Please refer to other texts for a full description of the various types of VSDs.

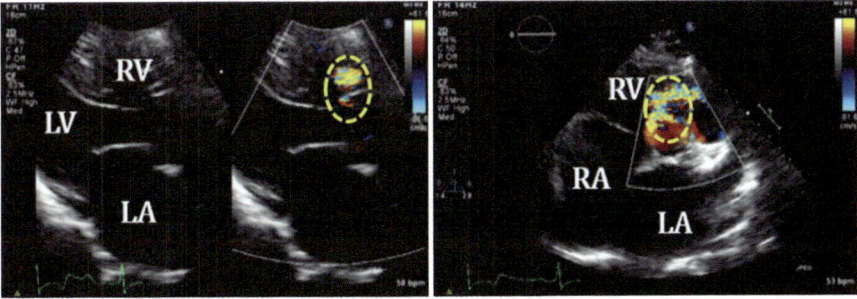

Fig. 12.9 Supracristal VSD. Parasternal long-axis view (left panel) and parasternal short-axis view (right panel) showing a supracristal VSD (*yellow oval*) with left to right flow

12.4.2 Quantifying the Degree of Shunt

- The shunt could not be accurately quantified on the echocardiogram for several reasons.
  - Very turbulent flow noted across the RVOT such that the RVOT pulse-Doppler signal showed aliasing and accurate measurement could not be obtained.
  - No view could be found where the VSD diameter could be reliably measured.

12.4.3 VSD Flow and Intracardiac Pressures

- The flow across the VSD showed the typical high velocity, systolic only flow (Fig. 12.10); peak systolic velocity was 5.28 m/sec corresponding to systolic ΔP_{LV-RV} of 112 mmHg.
- The patient's blood pressure at the time of the echocardiogram was 114/65 mmHg.
- Assuming LV systolic pressure = aortic systolic pressure → calculated RV systolic pressure based on the VSD flow velocity:
  - RV systolic pressure = LV systolic pressure − ΔP_{LV-RV}=114 − 112 = **2 mmHg**.
- The patient also had tricuspid regurgitation; peak TR velocity was 3.6 → systolic ΔP_{RV-RA} = 52 mmHg.
- The IVC was small with normal inspiratory collapse, suggesting normal RAP ~2–5 mmHg.
- Using the TR peak velocity and the estimated RAP → calculated RV systolic pressure
  - RV systolic pressure = TR peak gradient + RAP = 52 + 3 = **55 mmHg**.

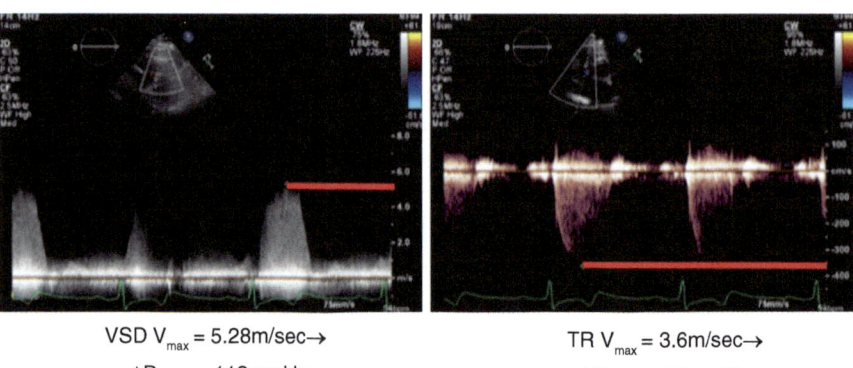

VSD V_{max} = 5.28m/sec→   TR V_{max} = 3.6m/sec→
ΔP_{LV-RV} = 112mmHg ΔP_{RV-RV} = 52mmHg

Fig. 12.10 Supracristal VSD and tricuspid regurgitation (TR) spectral Doppler. Left panel shows the Spectral Doppler of the VSD flow from the third patient presented. High-velocity systolic flow noted, with peak velocity 5.28 m/sec, corresponding to a systolic pressure gradient between the left ventricle and right ventricle (ΔP_{LV-RV}) of 112 mmHg. Right panel shows the tricuspid regurgitation (TR) spectral Doppler with peak TR velocity of 3.6 m/sec, corresponding to a systolic pressure gradient between the right ventricle and right atrium (ΔP_{RV-RA}) of 52 mmHg

12.4 Case Presentation

- Thus, interrogating the VSD flow and the TR flow led to contradicting conclusions.
- When the data obtained does not make sense, an explanation must be found:
  - Technical issues should be ruled out (e.g., measuring the wrong flow, nonparallel Doppler acquisition, etc.).
  - However, if no technical errors occurred (as in this case), a physiologic explanation must be found.
  - All assumptions that were used in order to draw the mismatched conclusions must be examined and verified or refuted.

    Assumption 1: LV systolic pressure equals the aortic systolic pressure. While this is true in many cases, there are instances where this is not the case. For example, in aortic stenosis, there is a gradient across the aortic valve and the LV pressure is higher than aortic pressure. Any obstruction to flow between the LV and the site of blood pressure measurement will cause the LV systolic pressure to be unequal and higher than aortic systolic pressure (see Chap. 6). In the case presented, no gradient was found anywhere across the path between the LV and measured BP.

    Assumption 2: The excessive turbulence seen in the RVOT was due only to the increased flow in the RV from the VSD. While it is common to see high flow in the RVOT in patients with VSD, there may be other explanations as well. VSD can be part of complex congenital anomaly and there may be additional pathologies. In the case presented, careful evaluation with spectral Doppler identified an additional area of flow acceleration **within** the right ventricle (Fig. 12.11). A systolic flow was seen in the proximal RVOT (flowing away from the transducer) with a peak velocity/gradient of approximately 4 m/sec, 64 mmHg.

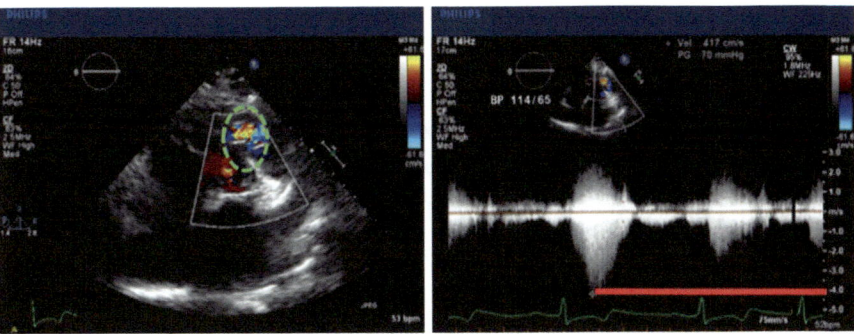

RV obs V_{max} = 4m/sec →

ΔP_{RV-RV} = 64mmHg

Fig. 12.11 Additional flow in the right ventricle. Careful probe manipulation and Doppler interrogation showing an additional area of flow acceleration within the right ventricle (*green oval*, left panel). Spectral Doppler (right panel) confirmed the systolic flow in the proximal RVOT, flowing away from the transducer, with a peak velocity/gradient of approximately 4 m/sec, 64 mmHg

- Thus, the assumption that the right ventricle is a single chamber with a single flow velocity across it appears to be mistaken.

12.4.4 Summary of VSD with RVOT Obstruction Case

- A patient with known congenital VSD was referred for an echocardiogram to assess pulmonary artery pressure.
- Mismatched data were obtained by interrogating the VSD flow and the TR flow.
- The VSD flow (high-velocity systolic flow) suggested normal pressure in the right ventricle while the high peak TR velocity suggested elevated RV pressure.
- An additional area of flow acceleration was found within the right ventricle proximal to the RVOT.
- Putting the data together (Fig. 12.12):
  - A supracristal ventricular septal defect is present.

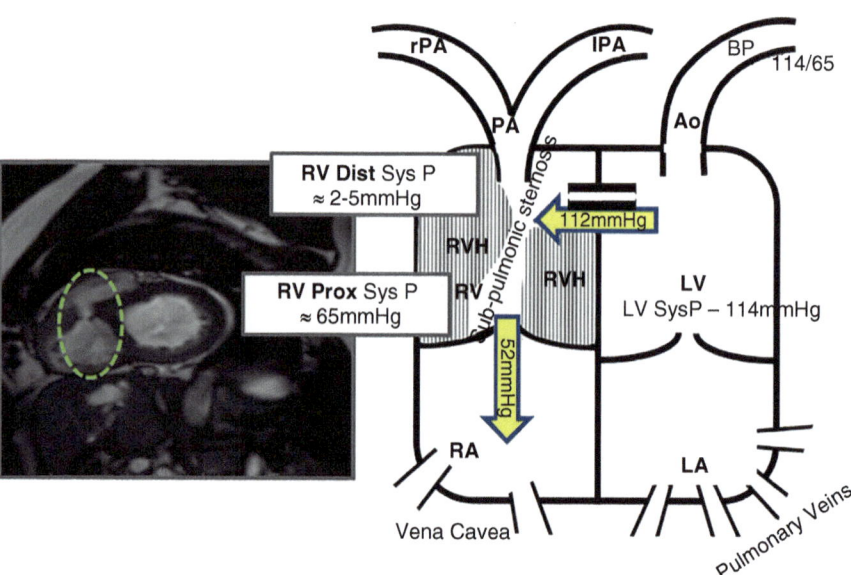

Fig. 12.12 MRI and diagram showing double-chamber right ventricle (RV). The patient had a supracristal VSD and functional sub-pulmonic stenosis (likely due to severe right ventricular hypertrophy [RVH]). The area of narrowing of the right ventricle can be clearly seen on the MRI (left panel, *green oval*). The RV essentially had two compartments: proximal inflow chamber, from which the tricuspid regurgitation (TR) originated. This part of the RV had increased systolic pressure, giving rise to the high-velocity TR that was seen. The outflow part of the RV was distal to the sub-pulmonic stenosis; in this distal compartment, the RV systolic pressure was low. The supracristal VSD opened into the distal, low-pressure RV chamber, such that VSD systolic gradient/velocity were high

- In addition, proximal RVOT obstruction is present creating a double-chamber RV.
- The VSD opens into the distal part of the RV / RVOT (beyond the RV obstruction), where the RV pressure is low. Thus, the VSD systolic velocity is high.
- The TR originates from the proximal (inflow) part of the RV, which is the pre-obstruction part of the RV. This part of the RV generates high pressure to overcome the RVOT obstruction.
- Thus, the TR velocity is high, representing proximal-RV systolic hypertension, however, there is **no** pulmonary hypertension.

12.5 Summary and Final Points

- Full evaluation of all available parameters is critical for full understanding of the impact of VSD on the heart.
- Careful attention to details is essential to avoid any technical errors that may confound the data.
- Interrogating the VSD flow can provide important hemodynamic information and data regarding intracardiac pressures.
- All the flows present should be interrogated and measured and should "make sense" when evaluated together; discrepancy in data obtained should prompt a thorough evaluation in order to obtain an understanding of the nature of the discordant data.

The Machinery Confusion

13

Abstract

Comparing the right-sided cardiac output and left-sided cardiac output is important in situations where a shunt is suspected; it is done by comparing the flow across the LVOT and the RVOT. It is important to remember the limitations and the pitfalls of this technique, as well as to understand precisely the information obtained from these calculations. PDA is a unique example where the classic comparison of the LVOT and RVOT cardiac output results in "confusing" data. Since PDA is an abnormal communication between the right and the left heart occurring outside the heart, it is important to understand which site (LVOT / RVOT) represents which circulation.

Keywords

Patent ductus arteriosus · Continuous flow · Qp:Qs ratio

Introduction

- Comparing the right-sided cardiac output and left-sided cardiac output is important in situations where a shunt is suspected.
- As described in previous chapters, this is done by comparing the flow across the LVOT and the RVOT.
- It is important to remember the limitations and the pitfalls of this technique, as well as to understand precisely the information obtained from these calculations.
- In this chapter, we will review a case that demonstrates the importance of understanding the circulation in the presence of a shunt, and how it relates to calculation of the shunt.

13.1 Case Presentation

- A 29-year-old man was referred for an echocardiogram for evaluation of a murmur.
- The murmur was described as a continuous, machinery-like murmur.
- There was no other significant past medical history; pediatric records were unobtainable.
- The echocardiogram showed continuous (systolic and diastolic) flow into the pulmonary artery, near the bifurcation of the main pulmonary artery.
- Spectral Doppler confirmed the presence of continuous high-velocity flow (Fig. 13.1).
- These findings were consistent with the presence of patent ductus arteriosus (PDA).

13.1.1 Approach to PDA Evaluation

- PDA is a congenital anomaly where the ductus fails to constrict and obliterate post-partum.
- During fetal life, the ductus allows blood to bypass the lungs and flow from the pulmonary artery into the aorta; after birth, pulmonary vascular resistance

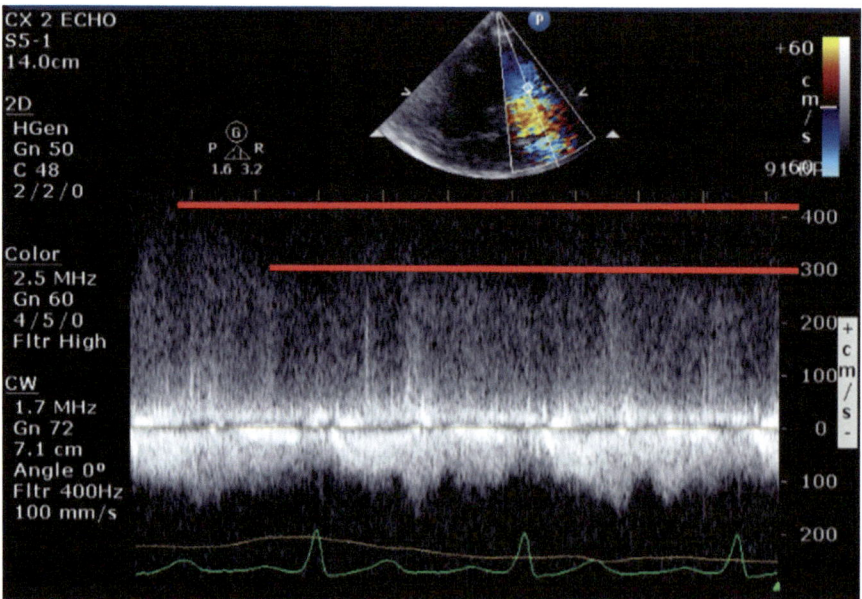

Fig. 13.1 Spectral Doppler of patent ductus arteriosus (PDA). The classic appearance of flow across the PDA is continuous, high-velocity flow during systole and diastole

13.1 Case Presentation

decreases and systemic vascular resistance increases and the ductus obliterates and closes and becomes the ligamentum arteriosum.
- In certain cases, the ductus does not obliterate and communication between the aorta and the pulmonary artery remains even after the newborn period.
- However, since the PVR decreases and the SVR increases, the flow in the ductus reverses: from the aorta to the pulmonary artery (unless pulmonary hypertension develops and PVR increases, at which point the shunt may reverse again).
- When PDA is found as an incidental finding or due to a murmur (without any associated symptoms), it is typically a small PDA, with a Qp:Qs ratio of less than 1.5:1.
- Unique echocardiographic characteristics of PDA:
  - Spectral Doppler pattern.
  - Confusion in calculating Qp:Qs.

Spectral Doppler Pattern in PDA

- As mentioned, PDA is a communication between the aorta and the pulmonary artery.
- When the PDA is small or moderate and the pulmonary vascular resistance is normal, the flow in the PDA is left to right—from the aorta to the pulmonary artery.
- The driving force for the PDA flow is the pressure gradient between the aorta and the pulmonary artery ($\Delta P_{Ao\text{-}PA}$).
- Calculating the driving force during systole and diastole:

	Ao (mmHg)	PA (mmHg)	ΔP (mmHg)	V (m/sec)
Systole	110	20	90	~4.7
Diastole	70	10	60	~3.8

- From the above: $\Delta P_{Ao\text{-}PA}$ is high both during systole and diastole → high-velocity flow during systole and diastole.
- Note: High-velocity diastolic flow is a unique finding; the only chamber that has high diastolic pressure is the aorta. Whenever a high-velocity diastolic flow is found, communication with the aorta should be looked for.

Confusion in Calculating Qp:Qs
- Estimating the magnitude of the shunt is an important part of PDA evaluation.
- Qp:Qs is calculated in the same way as described previously for ASD and VSD by comparing the pulmonary cardiac output and the systemic cardiac output (Figs. 13.2 and 13.3).
  - Note that the presence of valvular regurgitation (e.g., significant AI or PI) is a confounding factor that can cause an increase in the calculated local stroke volume (since it will include the regurgitant volume) → Qp:Qs cannot be estimated at the presence of significant valvular regurgitation.
- For the patient presented, no significant valvular regurgitation was seen.

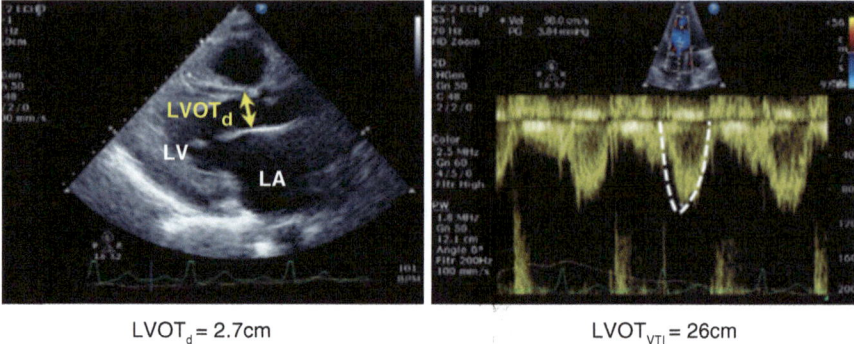

LVOT$_d$ = 2.7cm   LVOT$_{VTI}$ = 26cm

SV$_{left}$ = CSA X VTI = 5.7 × 26 = 149cc

Fig. 13.2 Qp:Qs calculation—left-sided stroke volume (SV$_{left}$). The LVOT diameter (LVOT$_d$) is measured from the parasternal long-axis view; the LVOT$_{VTI}$ is measured from an apical view. SV$_{left}$ is calculated as LVOT Cross-sectional area (LVOT CSA) x LVOT$_{VTI}$ (LA—left atrium, LV—left ventricle)

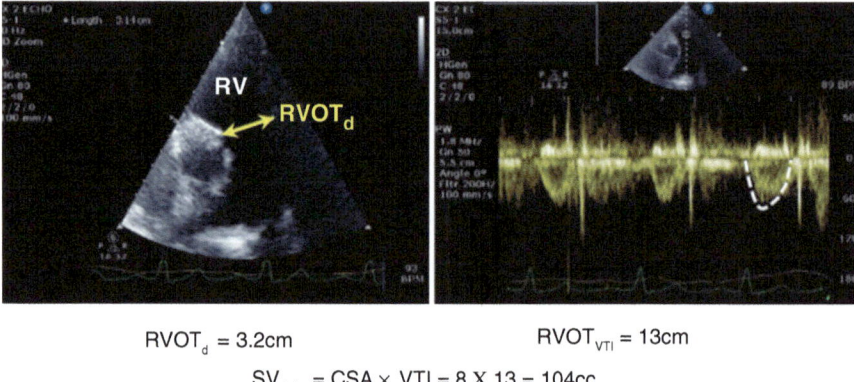

RVOT$_d$ = 3.2cm   RVOT$_{VTI}$ = 13cm

SV$_{right}$ = CSA × VTI = 8 X 13 = 104cc

Fig. 13.3 Qp:Qs calculation—right-sided stroke volume (SV$_{right}$). The RVOT diameter (RVOT$_d$) is measured from the parasternal short-axis view at the aortic valve level; the RVOT$_{VTI}$ is measured from the same view. SV$_{right}$ is calculated as RVOT Cross-sectional area (RVOT CSA) x RVOT$_{VTI}$ (RV—right ventricle)

- The following data were obtained:

	RVOT	LVOT
Diameter	3.2 cm	2.7 cm
VTI	13 cm	26 cm
Stroke volume	104 cc	149 cc

- The calculated LVOT stroke volume is **higher** than the RVOT stroke volume.

13.1 Case Presentation

- Evaluating the results:
  - In PDA (as opposed to ASD/VSD), the shunt occurs **outside** the heart.
  - Echo calculation of cardiac output is done at an **intracardiac** location—LVOT and RVOT.
  - If flow volume could be measured **distal** to the shunt (e.g., at the descending aorta and at the branch pulmonary arteries), it would become evident that the systemic circulation has a lower cardiac output than the pulmonary circulation.
  - Evaluating the PDA circulation (Fig. 13.4):

    Staring at the descending aorta (distal to the shunt)—flow volume is the effective stroke volume (SV_{eff}).

    SV_{eff} circulates throughout the body → capillaries → venules → IVC/SCV → right atrium → right ventricle → RVOT → main pulmonary artery.

    At the pulmonary artery, at the level of the shunt, shunt volume (V_{shunt}) is added to the flow volume such that the flow in the pulmonary artery is $SV_{eff} + V_{shunt}$.

    The increased volume ($SV_{eff} + V_{shunt}$) circulates from the pulmonary artery → pulmonary arterioles → pulmonary venules → pulmonary veins → left atrium → left ventricle → LVOT → proximal aortic root.

    From the above: at the RVOT, the volume calculated is the effective stroke volume—SV_{eff}, and at the LVOT the volume calculated is $SV_{eff} + V_{shunt}$.

    Yet, as shown, once the blood exits the heart, the shunt volume is added to the pulmonary flow volume.

    The pulmonary circulation sees the increased stroke volume; the systemic circulation sees only the effective stroke volume.

    Since the pulmonary circulation sees the increased stroke volume, which in turn flows back to the LA and LV, if the shunt is significant, left-sided volume overload develops.

    Going back to the patient's shunt calculation: the LVOT calculated stroke volume represents the $SV_{eff} + V_{shunt}$, meaning the Qp; the RVOT calculated stroke volume represents the SV_{eff}, meaning the Qs.

	RVOT	LVOT
Diameter	3.2 cm	2.7 cm
VTI	13 cm	26 cm
Stroke volume	104 cc (**Qs**)	149 cc (**Qp**)

Using the correct notation, the conclusion is: Qp:Qs = 1.4:1.

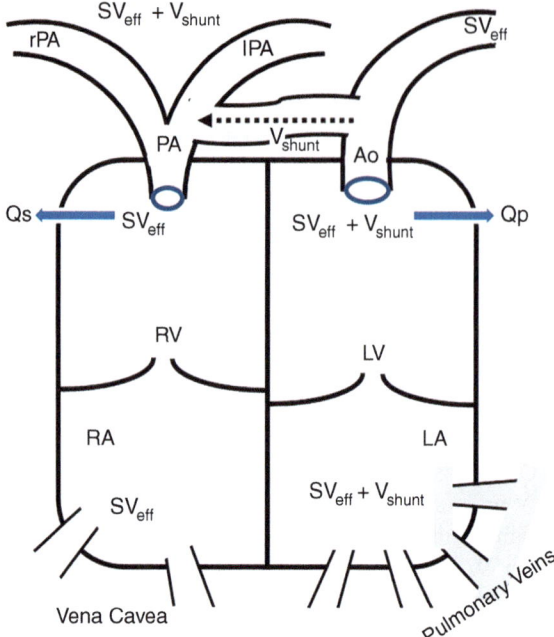

Fig. 13.4 Patent ductus arteriosus (PDA) circulation. The effective stroke volume (SV_{eff}), circulates through the systemic circulation, returns via the SVC/IVC to the right heart and flows through the right atrium (RA) and right ventricle (RV) and the right ventricular outflow tract (RVOT). At the pulmonary artery, at the level of the shunt, shunt volume (V_{shunt}) is added to the flow volume such that the flow in the pulmonary artery is $SV_{eff} + V_{shunt}$. The increased volume ($SV_{eff} + V_{shunt}$) circulates through the pulmonary circulation, returns via the pulmonary veins to the left heart, and flows through the left atrium (LA), left ventricle (LV), and left ventricular outflow tract (LVOT). Thus, the volume calculated at the RVOT is the effective stroke volume and the volume calculated at the LVOT is the $SV_{eff} + V_{shunt}$

13.2 Summary and Final Points

- PDA is an example of an abnormal communication involving the aorta, resulting in high-velocity diastolic flow in addition to high-velocity systolic flow.
- High-velocity diastolic flow is a unique finding since the only cardiac chamber with high diastolic pressure is the aorta.
- Identifying high-velocity diastolic flow should prompt a search for communication (congenital, iatrogenic) involving the aorta.
- Additionally, shunt quantification in PDA is an example for the importance of understating the circulation and how it effects echocardiographic shunt calculation.
- Comparing the flow across the LVOT and the flow across the RVOT is the mainstay of shunt quantification; it is important to understand which site represents which circulation (which may differ depending on the location of the shunt).

All Those AS Gradients

Abstract

Aortic stenosis (AS) is frequently encountered in echocardiography laboratories. Accurate evaluation of AS severity is essential for proper decisions regarding optimal treatment and timing of interventions. The severity of aortic stenosis is determined by the aortic valve area (at normal flow state), however, multiple criteria are utilized to grade AS severity: calculated aortic valve area, velocity/gradients across the valve, valvuloarterial impedance, valve calcifications, etc. The importance of accurately assessing AS severity and defining the subtype of AS cannot be overstated. It is essential to use meticulous techniques, including imaging from all views and utilizing the non-imaging transduced in order to measure maximal trans-aortic velocities. When discussing trans-aortic gradients it is necessary to specify exactly which gradient is reported—both for understanding the hemodynamic significance of the finding, as well as accurate communication when different techniques are utilized.

Keywords

Aortic stenosis · Maximal instantaneous gradient · Mean gradient · Peak-to-peak gradient

Introduction

- Aortic stenosis (AS) is frequently encountered in echocardiography laboratories.
- Accurate evaluation of AS severity is essential for proper decisions regarding optimal treatment and timing of interventions.
- The severity of aortic stenosis is determined by the aortic valve area (at normal flow state).

- Multiple criteria are utilized to grade AS severity: calculated aortic valve area, velocity/gradients across the valve, valvuloarterial impedance, valve calcifications, etc.).
- It is important to note that severe aortic stenosis can occur with severely elevated gradients across the valve, but also with gradients that may not appear severe; although valve area affects the gradients across the valve, it is not the only parameter that determines the gradients; other hemodynamic parameters that influence the flow across the valve impact the measured gradients.
- The importance of accurately assessing AS severity and defining the subtype of AS cannot be overstated.
- In this chapter, we will review the various gradients that can be measured by echocardiography and their relation to gradients measured by other techniques.
- For full description of evaluating AS and determining valve area and AS subtypes, please refer to other texts.

14.1 Technical Considerations

- Gradients across the aortic valve are measured by Doppler echocardiography.
- The Doppler equation describes the relation between the velocity of the target and the Doppler shift: $\Delta f = \dfrac{2f_0 V}{C} \times \cos\theta$, where θ is the angle of incidence between the measured flow and ultrasound beam (see Chap. 1).
- Rearranging the equation: $V = \dfrac{\Delta f \times C}{2f_0 (x \cos\theta)}$.
- Utilizing the Doppler equation allows calculation of the velocity of the interrogated flow (which is determined by the pressure gradient driving the flow—see Chap. 1).
- Note that the angle of incidence is important in the velocity calculation:
  - If a flow is interrogated at a parallel angle to the flow direction ($\theta = 0°$ or $180°$) → $\cos\theta = 1$.
- If, however, a flow is interrogated at a different angle, $\cos\theta < 1$.
- In clinical echocardiography the angle of incidence is not measured or estimated as this may introduce too much error; rather, an attempt is made to align the ultrasound beam as parallel as possible to the interrogated flow.
- The velocity calculation assumes a $\cos\theta = 1$ such that $V = \dfrac{\Delta f \times C}{2f_0}$.
- Thus, velocity as measured by Doppler evaluation, cannot be overestimated; if the angle of incidence is not parallel and angle correction is not used (as is the practice), the calculated velocity is underestimated relative to the true velocity.
- Hence, it is essential that the Doppler interrogation will be carried out at an angle of incidence as close as possible to a parallel angle to the examined flow.

- Given the significance of gradients evaluation in aortic stenosis assessment, it is of paramount importance that the flow across the valve is measured from multiple views and the highest obtained velocity is reported (assuming no physiologic beat-to-beat variability, as seen with atrial fibrillation for example).
- Every possible view should be used to measure the flow across the valve; right parasternal view (which is otherwise not a common view to use) must be included in AS evaluation; in elderly patients, more than 50% of the time the highest velocity may be obtained from the right parasternal view.
- Additionally, the non-imaging transducer should be routinely used when evaluating aortic stenosis; its smaller footprint allows manipulating it such that the Doppler beam may be better aligned with the aortic valve flow.

14.2 Maximal Instantaneous Gradient

- Measuring the peak velocity across the aortic valve allows calculation of maximal instantaneous gradient (MIG).
- Since continuous wave Doppler measures **all** the velocities along the path of the ultrasound beam, by using CW Doppler the highest velocity encountered can be measured.
- The MIG represents the point where the **gradient** (i.e., **difference**) between the LV pressure and the aortic pressure is highest; it does not necessarily occur when the LV or aortic pressure are at their highest (Fig. 14.1).

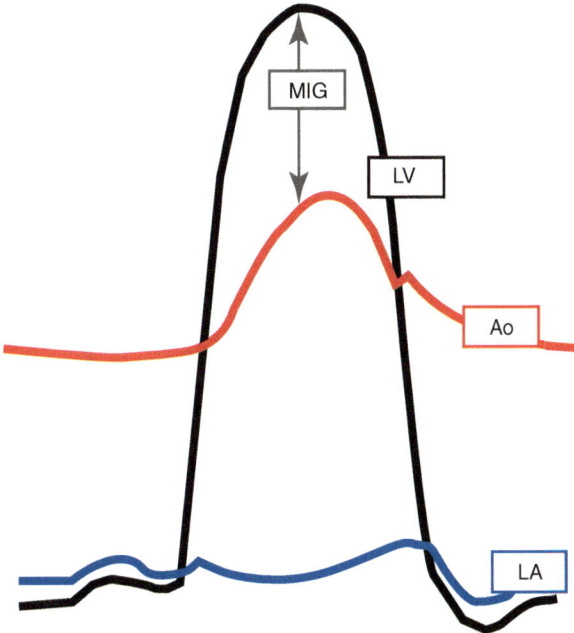

Fig. 14.1 Maximal instantaneous gradient (MIG). The MIG represents the point where the gradient (i.e., difference) between the left ventricular (LV) pressure and the aortic pressure (Ao) is highest; it does not necessarily occur when the LV or aortic pressure are at their highest

- In fact, in severe aortic stenosis, the aortic pressure rises slower than normal (the classic "pulsus tardus" felt on examination of the arterial pulse) reaching its peak later in systole, such that peak LV systolic pressure occurs earlier than peak aortic pressure.
- In order to measure the MIG by catheter-based technique, simultaneous measurement of LV and aortic pressures must be obtained by a dual-catheter technique.
- The MIG, by virtue of its definition, occurs for a brief period of time during systole; while it is an important parameter to measure and record as part of AS quantification, it does not fully represent the afterload against which the LV works.

14.3 Mean Gradient

- The mean gradient is an important parameter when assessing AS and considering its effect on the heart.
- As the name implies, it is the mean gradient that the LV "sees" throughout the course of systole.
- It is derived by tracing the continuous wave Doppler envelope; the computational software averages the derived pressure gradients over time and provides the mean gradient during the traced period (systole) (Fig. 14.2).

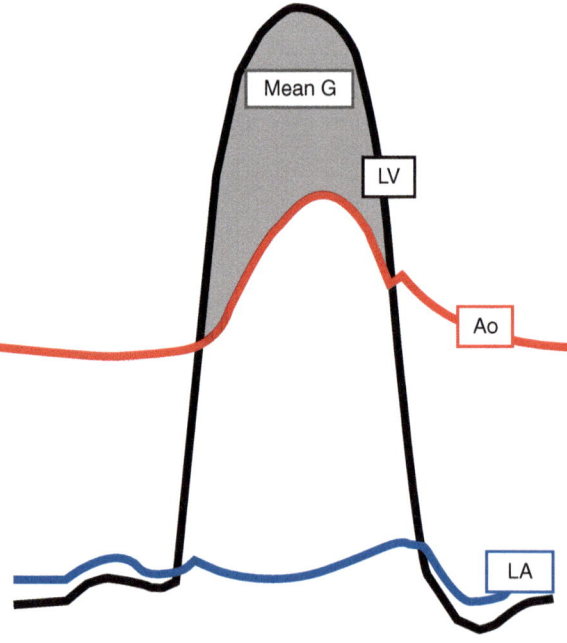

Fig. 14.2 Mean gradient. The mean gradient is what the LV "sees" throughout the course of systole. Tracing the continuous wave Doppler envelope allows the echo computational software to average the derived pressure gradients over time and to provide the mean gradient during the traced period (systole) (Ao—aorta, LA—left atrium, LV—left ventricle)

14.4 Peak-to-Peak Gradient

- Since mean gradient is calculated throughout systole, it is more closely related to the LV afterload.
- Mean gradient as obtained by Doppler echocardiography correlates very well with mean gradient as obtained by catheter-based measurement (utilizing dual-pressure gauge).

14.4 Peak-to-Peak Gradient

- Peak-to-peak gradient is a measurement obtained by catheter technique.
- It is acquired with a single gauge pressure catheter; a catheter is placed in the LV, recording the LV pressure.
- The catheter is pulled back into the ascending aorta and the aortic pressure is measured.
- The difference between the maximal LV pressure and the maximal aortic pressure is calculated (Fig. 14.3).
- It is noteworthy that peak-to-peak gradient is a measurement of a pressure gradient that does not actually occur at any point of the cardiac cycle.
- As discussed earlier, when aortic stenosis is present, the aortic pressure peaks late—with severe AS, the peak may be very close to the S2; at no point during systole is the LV pressure at peak and the aortic pressure at peak.
- Trans-aortic gradients obtained by Doppler echocardiography measure events that occur simultaneously and hence any Doppler based gradient is less comparable with the peak-to-peak gradient.
- If an echocardiographic estimation of peak-to-peak gradient is desired, it can be estimated using another technique:
  - If mitral regurgitation is present, the MR CW Doppler envelope can be used to measure the peak systolic gradient between the LV and the LA.
  - From above, LV peak systolic pressure = LA pressure + MR peak gradient.

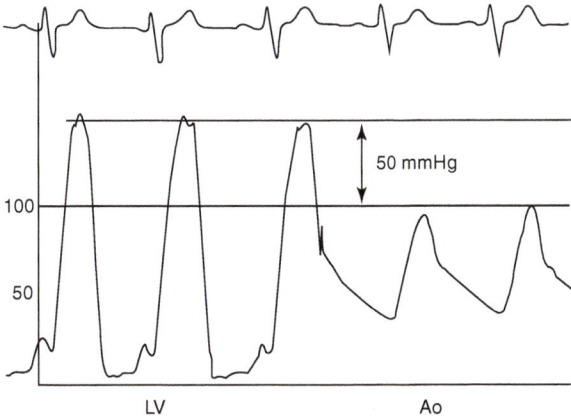

Fig. 14.3 Peak-to-peak gradient. This measurement is typically obtained by a catheter-based technique with a single gauge pressure catheter; a catheter is placed in the left ventricle (LV), then pulled back into the ascending aorta (Ao). The difference between the maximal LV pressure and the maximal aortic pressure is calculated

- LA pressure needs to be estimated based on clinical parameters and echocardiographic parameters (typically mitral inflow pattern, tissue Doppler of the mitral annulus, LA size, and estimated PA systolic pressure); in most cases, a conclusion of "normal" vs. "elevated" LAP can be reached → estimated peak LV systolic pressure can be calculated.
- Peak aortic pressure can be obtained by measuring the blood pressure.
- Subtracting aortic systolic pressure from estimated LV peak systolic pressure provides an estimate of the peak-to-peak gradient.

14.5 Summary and Final Points

- Aortic stenosis is a common diagnosis encountered in echocardiography laboratories.
- An important part of assessing AS severity is measuring the trans-aortic gradients.
- It is essential to use meticulous techniques, including imaging from all views and utilizing the non-imaging transduced in order to measure the maximal velocities.
- When discussing trans-aortic gradients it is necessary to specify exactly which gradient is reported—both for understanding the hemodynamic significance of the finding, as well as accurate communication when different techniques are utilized.

Dynamic LVOT 15

Abstract

Dynamic LVOT obstruction can be seen in many situations; examples include hypertrophic obstructive cardiomyopathy, Takotsubo cardiomyopathy, apical myocardial infarction with hypercontractile basal segments, inotropic therapy, and more. Significant LVOT obstruction may present acutely with shortness of breath and signs of hypoperfusion, mimicking the clinical picture of cardiogenic shock. Making the diagnosis of LVOT obstruction can be critical; therapies that are traditionally used for treating cardiogenic shock (i.e., inotropes, diuretics) can exacerbate and worsen the LVOT obstruction and a vicious cycle can be created. Echocardiography plays a central role in making this diagnosis. Provocation measures should be performed if no resting gradient can be found yet anatomic or clinical considerations suggest a possibility of inducible LVOT obstruction. All available data should be considered—m-mode, 2D imaging, color Doppler, and spectral Doppler. Careful attention to details is critical for demonstrating and grading the severity of LVOT obstruction.

Keywords

LVOT obstruction · Systolic anterior motion · SAM · Mitral regurgitation · High-velocity MR

Introduction

- Dynamic LVOT obstruction can be seen in many situations; examples include hypertrophic obstructive cardiomyopathy, Takotsubo cardiomyopathy, apical myocardial infarction with hypercontractile basal segments, inotropic therapy, and more.

- Significant LVOT obstruction may present acutely with shortness of breath and signs of hypoperfusion, mimicking the clinical picture of cardiogenic shock.
- Making the diagnosis of LVOT obstruction can be critical; therapies that are traditionally used for treating cardiogenic shock (i.e., inotropes, diuretics) can exacerbate and worsen the LVOT obstruction and a vicious cycle can be created.
- Echocardiography plays a central role in making this diagnosis.
- In this chapter, we will review the echo techniques that can be used to diagnose and quantify LVOT obstruction.

15.1 Anatomic Imaging

- Dynamic LVOT obstruction is due to systolic anterior motion (SAM) of the mitral valve.
- Structural findings that may contribute to the development of SAM include:
  - Narrow LVOT (e.g., due to septal hypertrophy).
  - Elongated / redundant anterior leaflet of the mitral valve.
  - Apical displacement of the papillary muscles.
- High-velocity blood flow enters the LVOT during systole, dragging the mitral leaflet toward the interventricular septum (the Venturi effect).
- The anterior mitral leaflet is usually involved in SAM, although cases of posterior leaflet SAM have been described.
- Presence and duration of leaflet-septal contact correlate with the severity of LVOT gradient.
- The systolic anterior motion can be identified on 2D imaging on the parasternal views or the apical views (Fig. 15.1).
- Anterior motion of the anterior mitral leaflet during diastole is a normal finding; it is common to see the anterior leaflet touching the septum during diastole and it should not be confused with SAM.
- Recognizing SAM often requires practice; it is typically an acquired "pattern recognition" skill.
- In addition, SAM of the mitral leaflet should be differentiated from chordal SAM; chordal SAM is common and rarely causes any hemodynamically significant obstruction.
- When identifying SAM of the mitral leaflet, it is useful to note if there is leaflet-septal contact, and if so, how long (relative to the length of systole) it lasts.
- M-mode (Fig. 15.2) can be particularly helpful in defining the extent and duration of SAM; often both the extent of the SAM and the duration can be easily recognized on m-mode tracing.
  - M-mode may also be helpful when looking for the classic mid-systolic closure of the aortic valve which is the echocardiographic correlate of pulsus bisferiens.
- SAM may not be present on resting echo, however, may be induced by various hemodynamic states or medications; high index of suspicion is required and

15.1 Anatomic Imaging

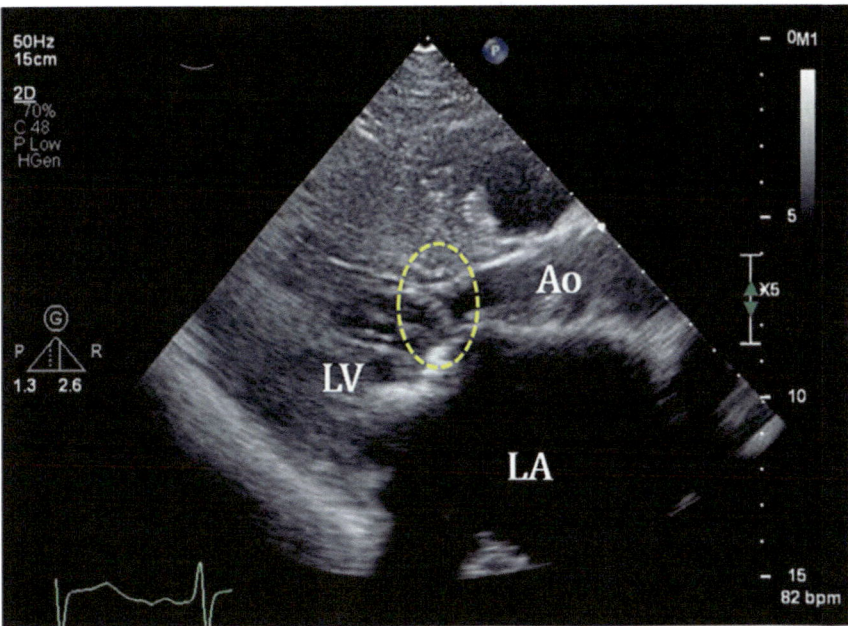

Fig. 15.1 Systolic anterior motion (SAM) of the mitral valve. Systolic frame from a parasternal long-axis view showing significant SAM; the anterior mitral leaflet reaches all the way to the interventricular septum (*yellow oval*) (Ao—aorta, LA—left atrium, LV—left ventricle)

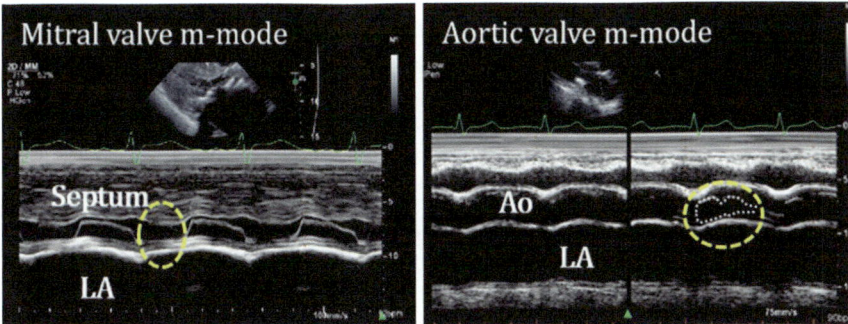

Fig. 15.2 M-Mode findings with SAM and LVOT obstruction. Mitral valve m-mode (*left panel*) can be very helpful in identifying systolic anterior motion (SAM) of the mitral valve. The systolic contact between the mitral leaflet and the septum can be clearly visualized (*yellow oval*). The duration of the leaflet-septal contact can also be recognized. M-mode of the aortic valve (right panel) can show mid-systolic closure of the aortic valve (*yellow oval*)

provocative maneuvers should be performed (e.g., Valsalva maneuver) when anatomic findings suggest a risk for provocable LVOT obstruction.
- In extreme situations, LVOT obstruction may be so dynamic that its presence may change from beat to beat (Fig. 15.3), even in the absence of significant change in the R-R interval.
- When provoking maneuvers are performed, it is advisable to assess **all** parameters (anatomic, color, and spectral Doppler) both at rest and with provocation.

15.2 Doppler Imaging

- Color Doppler imaging can provide clues to the presence of significant LVOT obstruction.
- Accurate quantifying of LVOT obstruction requires measurement of the LVOT gradient.

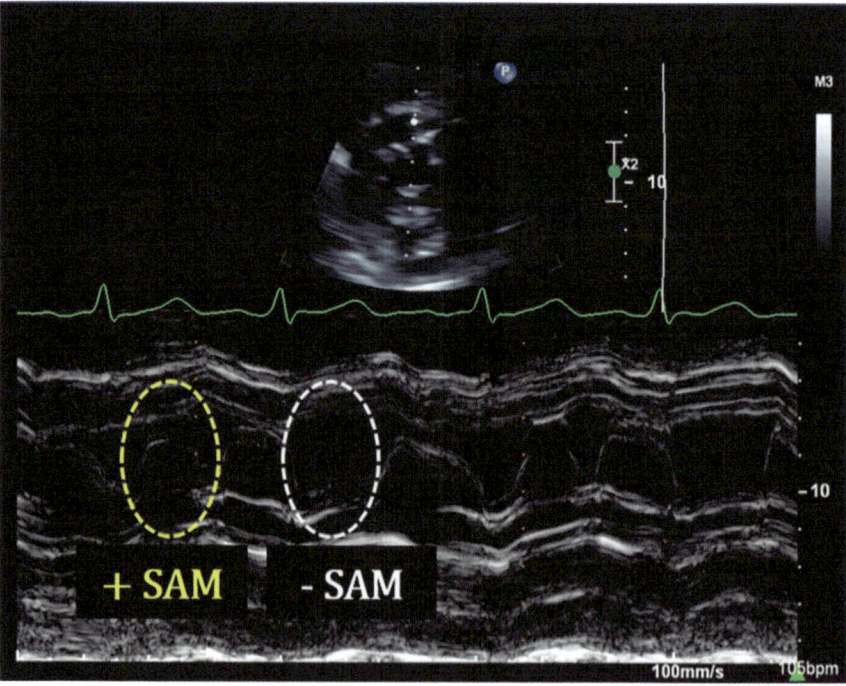

Fig. 15.3 Dynamically dynamic LVOT obstruction. Presence and duration of systolic anterior motion (SAM) of the mitral valve can change with varying hemodynamic conditions. In the mitral m-mode shown, extreme variability is seen, such that SAM is present every other beat; changes in preload and contractility create dynamic loading conditions that result in alternating presence of SAM

15.2.1 Color Doppler Imaging

- Color Doppler imaging does not allow accurate velocity measurements; only average velocities are displayed.
- By convention, flow toward the transducer is coded in red, away from the transducer in blue, and high-velocity flow (meaning average velocity above the Nyquist limit) is depicted in mosaic-yellow color.
- Typically, with LVOT obstruction, turbulent flow is seen in the LVOT.
- In addition, significant SAM of the mitral valve is often accompanied by significant mitral regurgitation; often, since the mechanism of the MR is the SAM, the mitral regurgitation jet is posteriorly directed.
- The classic appearance on color Doppler imaging shows the "crossed swords" sign: the turbulent LVOT jet perpendicular to the posteriorly directed turbulent MR jet (Fig. 15.4).
- Although this is a typical finding, it is noteworthy that the direction of the MR jet is not sufficient in and of itself to reliably discern whether independent mitral valve pathology is present in addition to the SAM; other techniques should be utilized in order to fully understand the valve anatomy and function.

15.2.2 Spectral Doppler Imaging

- Measuring the LVOT gradient is extremely important diagnostically and therapeutically.

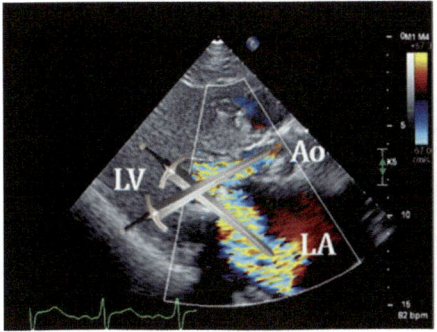
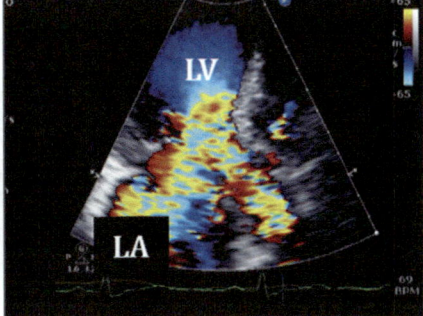

Fig. 15.4 Crossed swords sign. Parasternal long-axis view (left panel) and apical three chamber view (right panel) showing turbulent flow in the LVOT and severe mitral regurgitation (MR). When systolic anterior motion (SAM) of the mitral valve is present, often significant mitral regurgitation is seen. In these instances, classically the MR jet is directed posteriorly, such that the turbulent jet in the LVOT (from the LVOT obstruction) is perpendicular to the MR jet, creating the "crossed swords" sign. Yet, notably, the presence or absence of this sign is not sufficient evidence for ruling out other intrinsic mitral valve pathology as a cause for the MR (Ao—aorta, LA—left atrium, LV—left ventricle)

- Monitoring treatment success and decisions regarding invasive therapies rely on accurate assessment of the severity of the LVOT obstruction.
- Dynamic LVOT obstruction results in a typical spectral Doppler envelope: since the LVOT obstruction develops as systole is progressing, the obstruction is not present at the onset of systole but rather develops as systole evolves.
- This pattern is different than the one seen with aortic stenosis or with a fixed sub-aortic (LVOT) obstruction; in these instances, the obstruction is present from the "get go," at the onset of systole; the resulting spectral envelope is a mid-peaking, parabolic shaped spectral envelope (Fig. 15.5).
- Systolic cavity obliteration can also result in a systolic gradient; typically, cavity obliteration results in a very late-peaking jet (even later than LVOT obstruction jet). Additionally, intracavity gradient tends to be very short lived such that the classic appearance of the Doppler envelope is of a "stalactite" like jet.
- Although the shape and pattern of the spectral envelope can provide important clues as to the origin of the measured gradient, ultimately the most accurate echocardiographic way to localize the area of flow acceleration is by using pulse wave Doppler.
- Pulsing the LV-LVOT-AV tract can potentially localize the area of flow acceleration; unfortunately, when significant gradient is present, aliasing will occur and maximal velocity will not be discernable by pulse wave Doppler.
- Combining anatomic information with the CW-Doppler measured gradient commonly provides adequate information such that the presence and severity of LVOT obstruction can be diagnosed.
- If there are serial obstructions (i.e., cavity obliteration as well as dynamic LVOT obstruction, or LVOT obstruction and aortic stenosis) it may be challenging to separate the relative contribution of each pathology to the overall gradient.
- Occasionally, high PRF imaging can help with localization of a gradient.

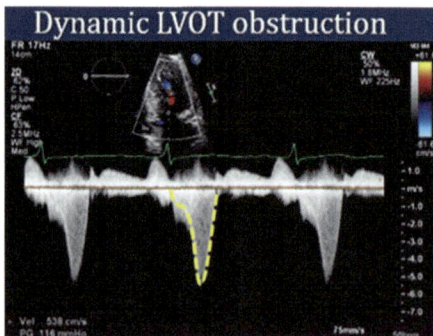
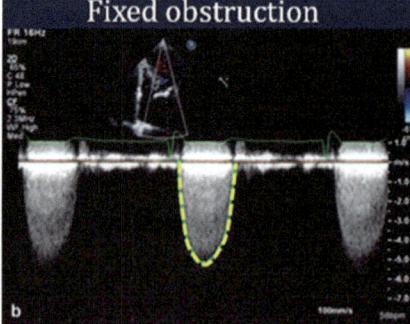

Fig. 15.5 Dynamic vs. fixed LVOT obstruction. Dynamic LVOT obstruction results in a typical spectral Doppler envelope (left panel). Since the LVOT obstruction develops as systole is progressing, the obstruction is not present at the onset of systole but rather develops late in systole. With aortic stenosis or with a fixed sub-aortic (LVOT) obstruction (right panel) the obstruction is present from the "get go," at the onset of systole such that the resulting spectral Doppler shows a mid-peaking, parabolic shaped envelope

15.2 Doppler Imaging

- When using high PRF imaging, signals from exact multiples of the distance to the first gauge are recorded.
- A recorded returning signal could originate at any site along the path of the beam that is located at a distance that is an exact multiple of the distance to the first sample volume.
- As the first (i.e., shallowest) sample volume is located very close to the transducer (determined by the high PRF utilized), the aliasing velocity is relatively high.
- This technique allows for partial range resolution.
- Combined with anatomic data obtained by the 2D imaging, it may help resolve a high-velocity flow along the LV-LVOT-AV tract.
• For all the techniques described (CW, PW, high PRF), care must be taken to try and optimize the view to verify that the peak velocity that is measured is in fact the flow being interrogated.
• However, occasionally it is difficult to separate the various flows encountered; in certain cases of severe MR with SAM and LVOT obstruction, it may be virtually impossible to obtain a "clean" envelops of the LVOT jet without catching any of the MR flow.
  - Of note, the MR Doppler can be useful in grading the severity of the LVOT obstruction:
  - The driving force for the MR is the systolic pressure difference between the LV and LA (ΔP_{LV-LA}).
  - When significant LVOT obstruction is present, the LV systolic pressure is elevated, making the ΔP_{LV-LA} higher than usual.
  - As discussed in Chaps. 6 and 7—typical MR peak velocity is approximately 5 m/sec.
  - When LV systolic pressure is elevated, MR peak velocity increases accordingly (Fig. 15.6).
  - If a full spectral envelope of the MR can be obtained, peak velocity and peak MR gradient can be calculated.
  - LV systolic pressure can be estimated as: LV sys = LAP + ΔP_{LV-LA}.
  - In most cases, normal vs. abnormal LAP can be discerned by echo thus allowing a rough estimation of the systolic LV pressure.
  - Measuring simultaneous blood pressure allows calculating the gradient between the LV and the aorta: gradient = LV sys Pressure − systolic BP.
  - If the anatomical data suggests LVOT obstruction, the above calculation provides a ball park estimate for the calculated LVOT gradient.
  - **Note**: When LVOT gradient and MR envelope can both be seen, **always** the MR jet is the one with the higher velocity; both flows are driven by the same systolic LV pressure, however, the LVOT flow velocity is determined by the gradient between the LV and the aorta and the MR velocity is determined by the gradient between the LV and the LA; the LA pressure is invariably lower than the aortic systolic pressure, hence the **MR gradient is invariably higher** than the LVOT gradient.

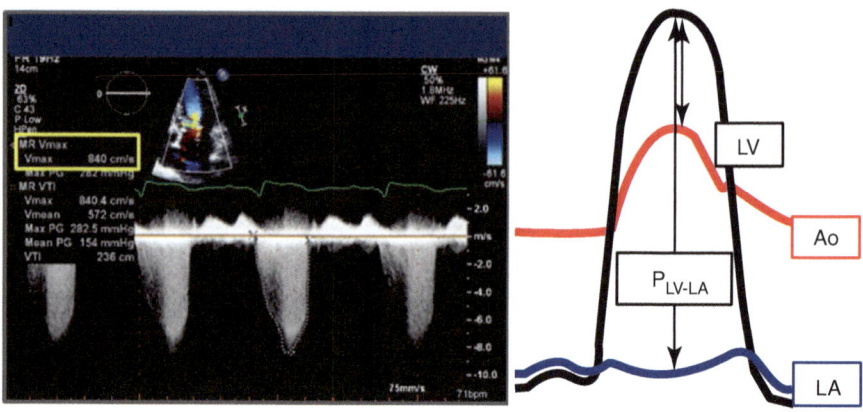

Fig. 15.6 Mitral regurgitation in LVOT obstruction. When LVOT obstruction is present, left ventricular (LV) systolic pressure is higher than aortic pressure (Ao). The high LV systolic pressure results in a very high systolic gradient between the LV and the left atrium (ΔP_{LV-LA}). The high ΔP_{LV-LA} creates a very high velocity MR jet (in the case shown, peak MR velocity 8.4 m/sec)

15.3 Summary and Final Points

- Dynamic LVOT obstruction is an important diagnosis to make since treatment of the condition is unique and different from other causes of cardiogenic shock and pulmonary edema.
- All available data should be considered—m-mode, 2D imaging, color Doppler, and spectral Doppler.
- Careful attention to details is critical for demonstrating and grading severity of SAM and LVOT obstruction.
- Provocation measures should be performed if low or no resting gradient can be found yet anatomic or clinical considerations suggest a possibility of inducible LVOT obstruction.

The Futile Heart

16

Abstract

In complex cases (i.e., aorto-right atrial fistula), in order to obtain a complete understating of the clinical picture and possibly recommend appropriate treatments, a thorough investigation is required with attention to technical details that can influence the various calculations and quantifications utilized. Combining all available echocardiographic data—m-mode, 2D, Doppler, 3D—can prove invaluable for understanding the hemodynamics in certain situations. Detailed echocardiographic evaluation can provide important information and be key to making a clinically important diagnosis. Any calculation utilized should be thought of and the information obtained from it should be understood in the larger context, and interpreted accordingly.

A significant aorta to right atrium fistula can be thought of as the "futile heart"—all four chambers and pulmonary circulation are exposed to volume overload while the systemic circulation is at a low cardiac output state. Echocardiography can identify the presence and severity of such a condition and help guide appropriate treatment.

Keywords

Endocarditis · Anasarca · Aorta-RA fistula

Introduction

- Combining all available echocardiographic data—m-mode, 2D, Doppler, 3D—can prove invaluable for understanding the hemodynamics in certain situations.
- In complex cases, in order to obtain a complete understating of the clinical picture and possibly recommend appropriate treatments, a thorough investigation is required with attention to technical details that can influence the various calculations and quantifications utilized.

16.1 Case Presentation

- A 50-year-old woman presented for evaluation of anasarca and increasing dyspnea; symptoms started subacutely with several weeks of peripheral edema, ascites, shortness of breath, and decreasing effort tolerance.
- Past medical history significant for mitral and aortic valve endocarditis approximately 9 months prior to presentation.
- At that time underwent aortic and mitral valve replacement as well as permanent pacemaker implantation.
- No records from her postoperative echocardiograms were available.
- Her echocardiogram upon presentation was significant for (Fig. 16.1).
    - Mild biventricular dilatation with mildly reduced LV and RV systolic function.
    - Normal appearing mitral and aortic bioprosthetic valves; thin leaflets, normal opening excursion, and no significant regurgitation.
    - Pacemaker wires in the RV and RA.
    - Dilated and plethoric IVC.
    - Tricuspid regurgitation seen; difficult to quantify; there appeared to be additional turbulent flow into the right atrium.
    - Spectral Doppler interrogation showed a continuous high-velocity flow into the RA (Fig. 16.2).

16.1.1 Unusual Flow into RA

- The most common turbulent flow into the RA is tricuspid regurgitation (TR).
    - Trivial-to-mild TR is commonly seen in healthy individuals.
    - More significant TR is abnormal and an attempt should be made to understand the etiology of the valve disease.
    - Importantly, TR is a systolic-only phenomenon; during diastole the tricuspid valve is open and blood flows from the right atrium into the right ventricle.

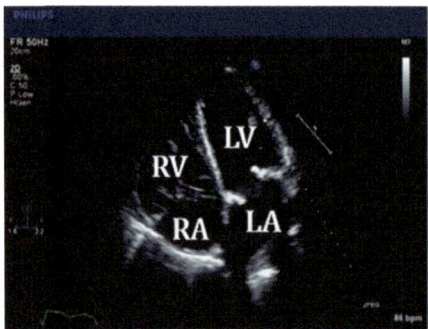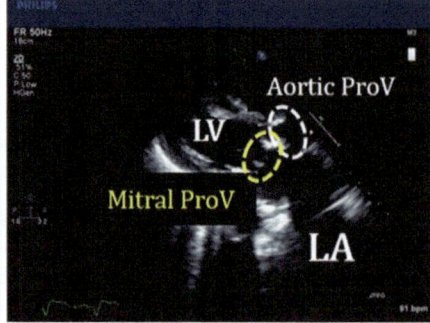

Fig. 16.1 Baseline echocardiogram. Apical four chamber view (*left panel*) showing mild biventricular dilatation. Parasternal long-axis view (*right panel*) showing normal appearing aortic bioprosthetic valve (Aortic ProV) and mitral bioprosthetic valve (mitral ProV) (LA—left atrium, LV—left ventricle, RA—right atrium, RV—right ventricle)

16.1 Case Presentation

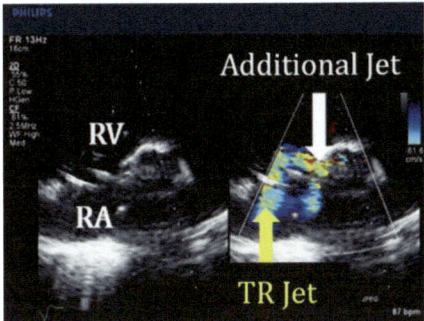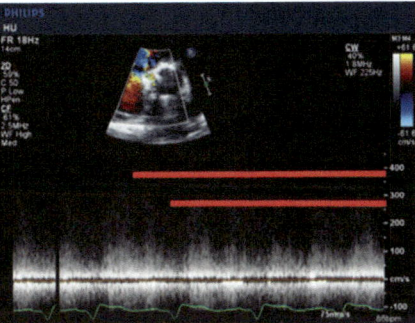

Fig. 16.2 Color and spectral Doppler of right atrial (RA) flow. Left panel showing color compare image from the parasternal short-axis view at the aortic valve level. Tricuspid regurgitation (TR) is seen with an additional, turbulent jet entering the right atrium. Right panel showing spectral Doppler of the "additional jet"; high velocity systolic and diastolic flow can be appreciated

- In the case presented, turbulent continuous flow (systolic and diastolic) was seen, and by color Doppler appeared more than mild.
- Continuous flow into the RA means that the flow originates from a chamber at which both systolic and diastolic pressures are higher than right atrial pressure (and the chamber is at proximity to the right atrium).
- Possibilities include: left ventricle or aorta.
    - Left ventricle to RA communication ("Gerbode" defect).
        May be congenital or acquired.
        Can be the result of a defect in the ventriculo-atrial septum or the result of a VSD combined with tricuspid leaflet perforation.
        The driving force for the flow across a Gerbode defect is the pressure gradient between the LV and the RA (ΔP_{LV-RA}).
        During systole: LV pressure ~ 120 mmHg, RA pressure ~ 5 mmHg → ΔP_{LV-RA} = 115 mmHg which correlates with high-velocity flow (~5 m/sec).
        During diastole: LV pressure ~ 10 mmHg, RA pressure ~ 5 mmHg → ΔP_{LV-RA} = 5 mmHg, which correlates with low-velocity flow (1 m/sec).
        In certain cases, diastolic RA pressure may be higher, even higher than LV diastolic pressure, resulting in absent or even reversed diastolic flow.
        In the case presented, the diastolic velocity was also high (nearly 3 m/sec) making it unlikely to have originated in the left ventricle.
    - Aorta to RA communication.
        Communication between the aortic root and right atrium can be iatrogenic or a complication of endocarditis with fistula formation.
        The driving force for the flow of an Ao-RA fistula is the pressure gradient between the aorta and the RA (ΔP_{Ao-RA}).
        During systole: Ao pressure ~ 120 mmHg, RA pressure ~ 5 mmHg → ΔP_{Ao-RA} = 115 mmHg, which correlates with high-velocity flow (~5 m/sec).
        During diastole: Ao pressure ~ 70 mmHg, RA pressure ~ 5 mmHg → ΔP_{Ao-RA} = 65 mmHg, which correlates with high-velocity flow (4 m/sec), although not as high as the systolic velocity.

As mentioned previously (Chap. 13), the **only** chamber that has high diastolic pressure is the aorta; whenever a high-velocity diastolic flow is seen, a search for communication with the aorta should be carried out.
- In the case presented, combining the color Doppler data (turbulent, at least moderate flow into the RA), with the spectral Doppler data (continuous high-velocity flow both in systole and diastole) leads to the conclusion that the patient had an aorta-right atrial fistula.
- The etiology of the fistula could be either from the initial endocarditis disease or a complication of the surgical intervention that was done to treat the infection.

16.1.2 Quantifying the Shunt

- The magnitude of the shunt between the aorta and the right atrium needs to be assessed.
- Importantly, the velocity of the shunt flow is **not** a manifestation of the magnitude of the shunt; as explained above, the velocity is determined by the pressure gradient driving the flow (which is a high-pressure gradient both during systole and diastole); the velocity would be the same whether there is a large shunt volume or a small shunt volume.
- Quantifying the shunt is important in order to decide if the shunt could be the cause of the clinical syndrome and thus might need to be fixed.
- Typically shunt can be quantified by echocardiography by directly measuring the flow across the communication or by calculating the Qp:Qs ratio.
    - Direct measurement of shunt flow:
        Direct measurement can be performed any place a spectral Doppler can be obtained and a cross-sectional area (CSA) can be measured.
        When these measurements are available, the flow volume is calculated as: volume = VTI × CSA.
        In our case neither one of these measurements could be reliably obtained: the flow velocity was too high and aliasing occurred on pulse wave Doppler; the fistula's diameter could not be visualized in such a way to obtain a reliable size measurement.
    - Calculating the Qp:Qs ratio
        Calculating the Qp:Qs ratio means comparing the flow across the pulmonary circulation and the systemic circulation.
        Typically this is done by measuring the diameters of the LVOT and RVOT and obtaining pulse wave Doppler VTI in these two areas.
        The flow volume in the RVOT is then compared to the flow volume in the LVOT and the ratio between them calculated.
        The difference between these volumes (assuming no significant valve regurgitation) is presumed to be due to the shunt.
        However, in the case presented, this technique would not be helpful due to the nature of the shunt (Fig. 16.3).

16.1 Case Presentation

Starting at the ascending aorta (distal to the shunt)—flow volume is the effective stroke volume (SV_{eff}).

SV_{eff} circulates throughout the body → capillaries → venules → IVC/SCV → right atrium.

At the right atrium, which is the level of the shunt, shunt volume (from the aortic root) enters the RA, such that the volume in the RA is $SV_{eff} + V_{shunt}$.

The increased volume ($SV_{eff} + V_{shunt}$) then flows across the tricuspid valve into the RV → RVOT → pulmonic valve → pulmonary circulation.

$SV_{eff} + V_{shunt}$ circulates through the pulmonary circulation → pulmonary capillaries → venules → pulmonary veins → left atrium.

From the left atrium, the increased volume ($SV_{eff} + V_{shunt}$) flows across the mitral valve into the LV → LVOT → aortic valve → aortic root.

At the aortic root, V_{shunt} flows into the RA and only the SV_{eff} continues to the systemic circulation.

From the above: both at the LVOT and at the RVOT, the flow volume is $SV_{eff} + V_{shunt}$.

Meaning that if Qp:Qs would be calculated by the classic method, the ratio that will be found is **1:1.**

Note that in the case of Ao-RA fistula, right heart chambers, the pulmonary circulation and left heart chambers are exposed to a volume overload, yet the systemic circulation does not see any of the increased volume.

If the shunt is significant, volume overload will develop on all four cardiac chambers and the pulmonary circulation, while the systemic circulation will suffer from a low cardiac output state.

This can be thought of as the "**ultimate futility**"; volume overload which the heart (all four chambers of it) needs to deal with, with no "benefit" systemically.

- Since the shunt cannot be quantified by the classic echocardiographic techniques, other assessments need to be done in order to estimate the significance of the finding.
- Given the description above, significant shunt would manifest as high flow at any intracardiac or proximal aortic location.
- In the case presented, the following data were obtained (Fig. 16.4):
  - **Aortic Valve**
    Peak velocity—3.3 m/sec
    Peak / mean gradient—43/26 mmHg
  - **Mitral Valve**
    Peak velocity—2.32 m/sec
    Peak / mean gradient—21/11 mmHg
    Pressure half time—94 mSec
  - **Left Main Coronary Artery**
    Peak velocity—1.2 m/sec

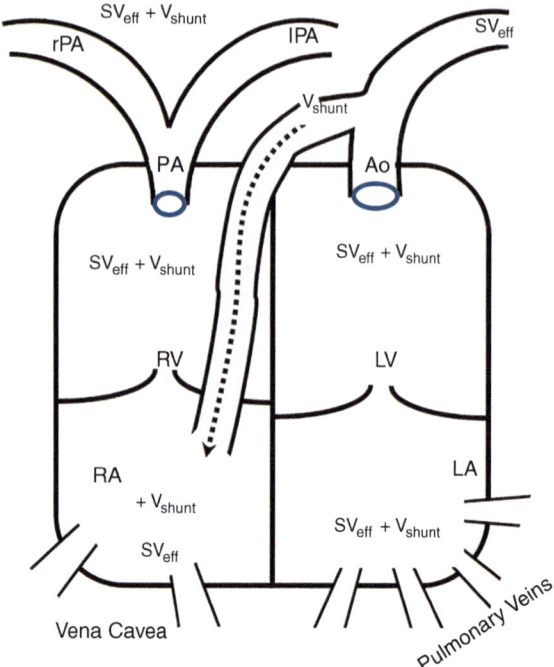

Fig. 16.3 Aorta to right atrium (Ao-RA) fistula circulation. The effective stroke volume (SV_{eff}), circulates through the systemic circulation and returns via the SVC/IVC to the right atrium (RA). At the RA, which is the level of the shunt, shunt volume (V_{shunt}) is added to the flow volume such that the flow volume becomes $SV_{eff} + V_{shunt}$. The increased flow volume circulates across the tricuspid valve, through the right ventricle (RV), the right ventricular outflow tract (RVOT), pulmonary artery (PA), and pulmonary circulation. $SV_{eff} + V_{shunt}$ then returns to the left heart via the pulmonary veins, and flows through the left atrium (LA), left ventricle (LV), left ventricular outflow tract (LVOT), and aortic root. At the level of the shunt (proximal ascending aorta), the shunt volumes flow into the RA; only the effective stroke volume SV_{eff} continues through the systemic circulation. Thus, volume calculated at the RVOT as well as the LVOT is the effective stroke plus the shunt volume ($SV_{eff} + V_{shunt}$)

- Normally, every prosthetic valve has some degree of flow acceleration/gradient.
- Normal prosthetic gradients vary by the type and size of the specific prosthesis; generally, for bioprosthetic aortic valve, velocity up to 2.5–3 m/sec can be normal; for mitral bioprosthetic a velocity of up to 1.9 m/sec can be normal.
- In the case presented, both trans-aortic and trans-mitral velocities were elevated, despite normal appearance of the valves, including on a TEE.
- High velocities despite normally functioning prosthetic valve are consistent with a high-flow state (in this case due to Ao-RA communication).
- In addition, the short pressure half time of the mitral inflow (94 mSec) is yet more evidence that the observed high gradient across the prosthesis is due to high flow rather than prosthetic stenosis.

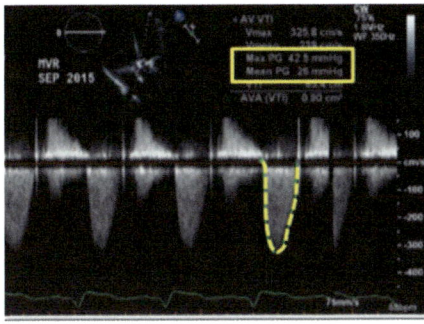 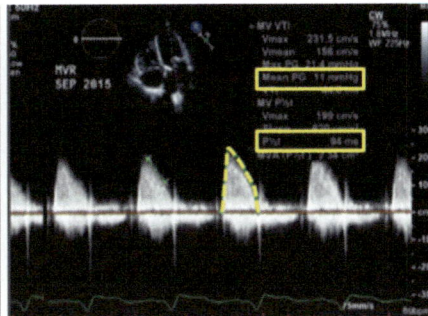

AV V$_{max}$ = 3.3m/sec
Peak/mean gradient = 43/26mmHg

Mean gradient = 11mmHg
PHT - 94mSec

Fig. 16.4 Evidence for significant shunt. Spectral Doppler across the aortic prosthetic valve (*left panel*) and mitral prosthetic valve (*right panel*). Both tracings show high velocity / gradients across the prosthetic valves, despite both valves appearing anatomically normal, with thin mobile leaflets and preserved opening (including by TEE evaluation)

- The flow in the left main coronary artery (by Doppler interrogation on TEE) was also measured to be elevated; the patient underwent a coronary angiogram a few days later and no obstructive disease was found; again, evidence of the high flow at the very proximal aortic root (proximal to the location of the fistula to the RA).
- Lastly, biventricular dilatation that was noted likely has also resulted from the volume overload state (although obviously, it is hard to definitively determine a cause and effect relationship).
- From all the above, despite inability to calculate precisely the shunt volume by echo, it appears that the Ao-Ra fistula created a hemodynamically significant shunt which was likely the cause of the patient's presenting symptoms.

16.2 Summary and Final Points

- Detailed echocardiographic evaluation can provide important information and be key to making a clinically important diagnosis.
- Any calculation utilized should be thought of and the information obtained from it should be understood; in the case presented, calculating Qp:Qs ratio of 1:1 should not have implied that the shunt is insignificant.
- A significant aorta to right atrium fistula is an example of a "futile heart"—all four chambers and pulmonary circulation are exposed to volume overload while the systemic circulation is at a low cardiac output state.
- On a final note, the patient ended up undergoing a percutaneous closure of the Ao-RA fistula; upon obliteration of the communication, velocities/gradients across the prosthetic valves, as well as the coronary flow velocities all, returned to normal. Clinically the patient did well with resolution of her symptoms.

Index

A
Anasarca, 138, 139
Aorta to right atrium (Ao-RA) fistula circulation, 142
Aortic insufficiency (AI), 79
   hemodynamic, 83–87
   parameters for, 82
   quantification, 80–83
Aortic stenosis (AS), 123
   gradients
      maximal instantaneous gradient (MIG), 125, 126
      mean gradient, 126, 127
      peak to peak gradient, 127, 128
      technical considerations, 124, 125
   severity, 123
Aortic valve, 141
Atrial septal defect (ASD), 90, 96
   anatomic type of, 90
   chamber size and function, 94, 97
   evaluation of, 90
   flow spectral Doppler, 91
   interatrial shunt, magnitude of, 92, 97
      Qp:Qs calculation, 92, 93
      shunt flow measurement, 93, 94
   pulmonary hypertension, 94, 95, 98
   pulmonary vascular resistance, estimation of, 95, 98, 99
   shunt direction, 91, 96

B
Bernoulli equation, 53, 61
Bernoulli principle, 3, 4
Border-tracing based technique, 12

C
Cardiac index, 77
Cardiac output (CO), 12, 77
Cardiogenic shock, 61, 64
Cardiomyopathy, 76
Cardiomyopathy with VSD
   anatomic type of, 108, 111
   flow and intracardiac pressures, 109, 110, 112, 113
   quantifying, degree of shunt, 108, 112
Color Doppler, 91
Color Doppler imaging, 66, 133
Congenital VSD, 103
Crossed swords sign, 133

D
Diastolic mitral regurgitation (MR)
   early, 71
   in first degree heart block, 66
   in high degree AV block, 70
Doppler echocardiography, 124, 127
Doppler effect, 2
Doppler shift, 2
Dynamic LVOT obstruction, 129, 130
   anatomic imaging, 130, 132
   Doppler imaging, 132
      color Doppler imaging, 133
      spectral Doppler imaging, 133–135
Dynamic *vs.* fixed LVOT obstruction, 134

© The Editor(s) (if applicable) and The Author(s), under exclusive license to Springer Nature Switzerland AG 2021
G. Perk, *Hemodynamics in the Echocardiography Laboratory*,
https://doi.org/10.1007/978-3-030-79994-6

E
Echocardiography, 1, 9, 10, 21, 27, 65
  Bernoulli principle, 3, 4
  continuity principles, 14, 15
  Doppler principle, 2, 3
  PISA calculation, 15–17
  pulmonary artery systolic pressure, 22
  resistance calculations, 13, 14
  TR spectral Doppler envelope, 22
  volume calculations
    border-tracing based technique, 12
    VTI based techniques, 10–12
  volumetric calculation, 18–20
  Wiggers diagram, 5, 6
Echo contrast, 12
Effective regurgitant orifice area
    (EROA), 15, 17
  anatomic/color assessment for, 40
  3D based direct measurement of, 40
Eisenmenger syndrome, 96
Ejection fraction, 12
Extreme pulses alternans, 76–78

F
First degree heart block, abnormal MR
    timing, 66–68
Futile heart
  shunt quantification, 140–143
  unusual flow into RA, 138–140

G
Garbode defect, 104, 139

H
High degree heart block, abnormal MR
    timing, 69, 71
High velocity mitral regurgitation (MR), 55
Hyperbolic paraboloid, 18

I
Inferior vena cava (IVC), 25
Intracardiac shunt, 101

L
Left main coronary artery, 143
Left subclavian artery Doppler, 56
Left ventricular outflow tract (LVOT), 120
  and aortic valve spectral Doppler, 56
  assessment, 77

Low velocity mitral regurgitation (MR), 60
LV outflow, 19
LVOT, *see* Left ventricular outflow tract

M
Machinery confusion, 117
Maximal instantaneous gradient (MIG),
    125, 126
Mean gradient, 126, 127
Mean pulmonary artery pressure, 29
  PA acceleration time, 31
  PI early diastolic velocity, 30
  on sys/dias pulmonary artery pressure, 30
  TR VTI, 31
Membranous VSD, 108
Method of discs, 13
Missing atrial contraction, 71, 72
Mitral regurgitation (MR), 18, 51, 52, 59
  abnormal MR timing
    first degree heart block, 66–68
    high degree heart block, 69, 71
    missing atrial contraction, 71, 72
  anatomic/color assessment for EROA, 40
  3D based direct measurement of, 40
  case study, 52, 60
  dP/dT calculation, 63
  driving pressures, 54
  general principles, 38–40
  gradient, 135
  in LVOT Obstruction, 136
  spectral Doppler, 53
  unusual MR tracing, 53–55
    abnormal envelope shape, 61–63
    low peak velocity, 61
Mitral valve, 141
  color Doppler, 87
  m-mode, 85
MR, *see* Mitral regurgitation

P
Patent ductus arteriosus (PDA)
  circulation, 122
  confusion in calculating Qp:Qs, 119, 121
  evaluation, 118
  spectral Doppler pattern, 119
Peak to peak gradient, 127, 128
Post infarct ventricular septal defect, 102
Proximal isovelocity surface area (PISA)
    calculation, 15–17
Pulmonary artery diastolic pressure, 28, 29
Pulmonary artery systolic pressure, 22
Pulmonary hypertension, 94, 95, 98

Index

Pulmonary outflow acceleration time (AT), 33
Pulmonary pressure
　data interpretation, 25, 26
　peak TR velocity measurement, 24
　right atrial pressure assessment, 25
　TR spectral Doppler envelope, 22, 24
Pulmonary vascular resistance (PVR), 13, 32
　estimation of, 95, 98, 99
　pressure drop and flow across pulmonary circulation, 34
　TR ΔP and RVOT flow, 32, 33
Pulmonic insufficiency (PI) spectral Doppler, 28
Pulmonic stenosis, 26

R
Right atrial pressure assessment, 25

S
Section surface area (SSA), 17
Secundum Atrial Septal Defect (ASD), 90
Secundum Atrial Septal Defect (ASD) With Eisenmenger syndrome, 96
Severe aortic stenosis, 124
Severe pulmonary hypertension, 30
Simpson's rule/method of discs, 12
Spectral Doppler, 10
Spectral Doppler imaging, 133–135
Spectral Doppler of patent ductus arteriosus (PDA), 118
Spectral Doppler pattern in PDA, 119
Stroke volume (SV), 12
Subclavian stenosis, 55
Supracristal VSD, 111
Systolic anterior motion (SAM), 130, 131
Systolic dysfunction, 76
Systolic pulmonary artery pressure, 24

T
Tissue Doppler, 76
Tissue Doppler tracing, 76
Transcutaneous aortic valve replacement (TAVR), 83
Tricuspid regurgitation (TR), 23, 138
　envelope, 32
　torrential, 23
　TR spectral Doppler envelope, 22, 24

V
Velocity time integral (VTI), 11
Ventricular septal defect (VSD), 101, 102
　anterior wall myocardial infarction
　　anatomic type of, 103
　　flow and intracardiac pressures, 104–107
　　quantifying degree of shunt, 103, 104
　cardiomyopathy case
　　anatomic type of, 108, 111
　　flow and intracardiac pressures, 109, 110, 112, 113
　　quantifying the degree of shunt, 108, 112
　evaluation of, 102
　post MI VSD case, 107
VTI based technique, 10–12

W
Wiggers diagram, 5, 6

GPSR Compliance

The European Union's (EU) General Product Safety Regulation (GPSR) is a set of rules that requires consumer products to be safe and our obligations to ensure this.

If you have any concerns about our products, you can contact us on ProductSafety@springernature.com

In case Publisher is established outside the EU, the EU authorized representative is:

Springer Nature Customer Service Center GmbH
Europaplatz 3
69115 Heidelberg, Germany

Batch number: 08823208

Printed by Printforce, the Netherlands